The Century of the Gene

# THE CENTURY OF THE GENE

EVELYN FOX KELLER

HARVARD UNIVERSITY PRESS
Cambridge, Massachusetts, and London, England

Printed in the United States of America

Third printing, 2002

First Harvard University Press paperback edition, 2002

Illustration on page vi by Nick Thorkelson

*Library of Congress Cataloging-in-Publication Data*
Keller, Evelyn Fox, 1936–
    The century of the gene / Evelyn Fox Keller.
       p. cm.
    Includes bibliographical references and index.
    ISBN 0-674-00372-1 (cloth)
    ISBN 0-674-00825-1 (pbk.)
    1. Genetics—History—20th century.   I. Title.
QH428 .K448 2000
576.5—dc21      00-038319

# CONTENTS

Introduction

THE LIFE OF A POWERFUL WORD

In 1900 three papers appeared in the same volume of the *Proceedings of the German Botanical Society*—the first by Hugo de Vries, the second by Carl Correns, and the third by Erich von Tschermak. De Vries, Correns, and Tschermak had independently "rediscovered" the rules of inheritance that Gregor Mendel, at the time an obscure Austrian monk, had found forty years earlier in his solitary investigations of pea plants. Mendel's original paper may have failed to attract much attention, but these papers did not. Indeed, they are generally credited not only with rescuing Mendel from oblivion but also with launching the science that would soon be called "genetics," and with that new science the age I am calling "the century of the gene."

The actual term *genetics* was coined in 1906, when William Bateson informed the International Congress of Botany that "a new and well developed branch of Physiology has been created. To this study we may give the title Genetics."[1] The term *gene* came along three years later, introduced by Wilhelm Johannsen. What was a gene? This no one could say. Johannsen himself wanted a new word so that it might be free of the taint of preformationism associated with such precursor terms as Darwin's *gemmules* (his units of "pangenesis"), Weismann's *determinants*, or de Vries' *pangens*. "Therefore," he wrote, "it appears simplest to isolate the last syllable, 'gene,' which alone is of interest to us . . . The word 'gene' is completely free from any hypotheses; it expresses only the evident fact that, in any case, many characteristics of the organism are specified in the gametes by means of special conditions, foundations, and determiners which are present in unique, separate, and thereby independent ways—in short, precisely what we wish to call genes."[2]

Two years later, Johannsen added, "The 'gene' is nothing but a very applicable little word, easily combined with others, and hence it may be useful as an expression for the 'unit factors,' 'elements' or 'allelomorphs' in the gametes, demonstrated by modern Mendelian researches . . . As to the nature of the 'genes,' it is as yet of no value to propose any hypothesis; but that the notion of the 'gene' covers a reality is evident in Mendelism."[3] A little word, perhaps—but a remarkably powerful one nonetheless. Indeed, this little word proved powerful enough to guide research in the science of genetics for the remainder of the century.

Not surprisingly, Johannsen's strictures against hypotheses about the material nature of the gene were rather less influential. As late as 1933, T. H. Morgan might claim, "There is no consensus opinion amongst geneticists as to what the genes are—whether they are real or purely fictitious."[4] Yet for the majority of Morgan's colleagues (indeed, for Morgan himself), genes had by then become incontrovertibly real, material entities—the biological analogue of the molecules and atoms of physical science, endowed with the properties that would make it possible, as de Vries had written, "to explain by their combinations the phenomena of the living world."[5]

For H. J. Muller, a student of Morgan's, the gene was not only "the fundamental unit of heredity" but "the basis of life."[6] Thus, for Muller, as for many other geneticists of the time, the question that begged was crucial: Just what sort of entity is a gene? Perhaps it was some sort of chemical molecule, but what sort? What is it made of, how big is it, and, above all, from whence comes its miraculous power to determine the properties of a developing organism and, at the same time, ensure the stability of those properties from one generation to another?

For the first four decades of this century, progress in genetics was steady and cumulative, but it offered little in the way of answers to such basic questions as these. The beginnings of an answer to the question of what genes are made of came in 1943 with Avery, MacLeod, and McCarty's identification of DNA as the carrier of biological specificity in bacteria. At roughly the same time, the first hint of what a gene does was provided by the "one gene—one enzyme" hypothesis of George Beadle and Edward Tatum. But it was the triumphal announcement by James D. Watson and Francis Crick in 1953 which convinced biologists not only that genes are real molecules but also that they are constituted of nothing more mysterious than deoxyribonucleic acid.[7] Thus, by midcentury, all remaining doubts about the material reality of the gene were dispelled and the way was cleared for the gene to become the foundational concept capable of unifying all of biology. Moreover, the identification of DNA as the genetic material spawned a new era of analysis, in which the powerful techniques of molecular genetics would replace those of classical genetics. As everyone knows, the ensuing progress has been spectacular, and it continues to accelerate.

In many ways, the advances of the last twenty-five years have been the most dramatic of the century (as well as the most publicized), and they have come largely as a consequence of, first, the advent of recombinant DNA technology in the mid 1970s and, second, the launching of the Human Genome Project (HGP) in 1990. Building on the phenomenal advances of molecular genetics, this enterprise—somewhat misleadingly named insofar as its mission has been to sequence not only the human genome but the genomes of other organisms of interest to biologists as well—has promised to reveal the genetic blueprint that tells us who we are. Indeed, it would be hard to imagine a more dramatic climax to the efforts of the entire century than the recent announcement that a draft of the entire sequence of the human genome will be completed in time to mark the centennial. At the very least, that announcement comes as a fitting climax to the career of the man who has been one of the prime movers behind this project: as Watson himself has put it, "Start out with the double helix and end up with the human genome."[8]

When the Human Genome Project was first proposed in the mid 1980s, it evoked a great deal of skepticism. But today, as its pace exceeds all expectations, few skeptics remain. So far, the complete genomes of over twenty-five microbial organisms have been sequenced, including those of that illustrious bacterium *Escherichia coli* on which molecular biology first cut its teeth. Genomes of more sophisticated model organisms have also been sequenced: yeast was the first, followed in 1998 by the roundworm *Caenorhabditis elegans*—

the first higher organism to be sequenced. The fruit fly *Drosophila*, the most famous of all model organisms in the history of genetics, made its debut in February 2000. The task of sequencing the human genome itself began relatively late, but its progress has been breathtaking: Less than 3 percent of the human genome had been sequenced by the end of 1997; by November 30, 1998, that number had risen to 7.1 percent; by September 5, 1999, it had reached 22 percent, and by the end of 1999, 47 percent. The expectation is that we should have a complete draft of the sequence of the human genome before the end of the year 2000.

I confess to having been one of the early critics. Like many others, I believed that so exclusive a focus on sequence information was both misguided and misleading. But today I am ready to share in the general enthusiasm for the HGP's achievements, although from a somewhat unusual perspective. What is most impressive to me is not so much the ways in which the genome project has fulfilled our expectations but the ways in which it has transformed them.

The aim of this book is to celebrate the surprising effects that the successes of this project have had on biological thought. Contrary to all expectations, instead of lending support to the familiar notions of genetic determinism that have acquired so powerful grip on the popular imagination, these successes pose critical challenges to such notions. Today, the prominence of genes in both the general media and the scientific press suggests that in this new science of genomics, twentieth-century genetics has achieved its apotheosis. Yet, the very successes that have so stirred our imagination have also radically undermined their core driving concept, the concept of the gene. As the HGP nears the realization of its goals, biologists have begun to recognize that those goals represent not an end but the beginning of a new era of biology. Craig Stephens writes, "Sequence gazing alone cannot predict with confidence the precise functions of the multitude of encoding regions in even a simple genome!" For this reason, he continues, "the era of genomic analysis represents a new beginning, not the beginning of the end, for experimental biology."[9]

To see how progress in genomics has begun to transform the way many biologists think about genes and genetics, and even about the meaning of the genome project itself, it is useful to recall the expectations with which that project began. A decade ago, many biologists spoke as if sequence information would, by itself, provide all that was necessary for an understanding of biological function. Spelling out his "Vision of the Grail," Walter Gilbert wrote, "Three billion bases of sequence can be put on a single compact disc (CD), and one will be able to pull a CD out of one's pocket and say, 'Here is a human being; it's me!'"[10] Today, almost no one would make such a provocative claim. Doubts about the adequacy of sequence information for an understanding of biological function have become ubiquitous, even among molecular biologists, and largely as a consequence of the increasing sophistication of genomic research. Instead of a "Rosetta Stone," molecular geneticist William Gelbart suggests that "it might be more appropriate to liken the human genome sequence to the Phaestos Disk: an as yet undeciphered set of glyphs from a Minoan palace . . . With regard to understanding the A's, T's. G's, and C's of genomic sequence, by and large, we are functional illiterates."[11]

Now that the genomes of several lower organisms have

been fully sequenced, the call for a new phase of genome analysis—*functional* genomics rather than structural genomics—is heard with growing frequency.[12] Hieter and Boguski define functional genomics as "the development and application of global (genome-wide or system-wide) experimental approaches to assess gene function by making use of the information and reagents provided by structural genomics."[13] In their view, the sequence no longer appears as an end-product but rather as a tool: "The recent completion of the genome sequence of the budding yeast . . . has provided the raw material to begin exploring the potential power of functional genomics approaches."[14] In a similar vein, anticipation of the full sequence of the *Drosophila* genome found geneticists who study this organism girding for a long haul. As Burtis and Hawley put it, they are preparing for "the huge amount of work that will be involved in correlating the primary DNA sequence with genetic function . . . This link is essential if we are to bring full biological relevance to the flood of raw data produced by this and other projects to sequence the genomes of 'model' organisms."[15]

It is a rare and wonderful moment when success teaches us humility, and this, I argue, is precisely the moment at which we find ourselves at the end of the twentieth century. Indeed, of all the benefits that genomics has bequeathed to us, this humility may ultimately prove to have been its greatest contribution. For almost fifty years, we lulled ourselves into believing that, in discovering the molecular basis of genetic information, we had found the "secret of life"; we were confident that if we could only decode the message in DNA's sequence of nucleotides, we would understand the "program" that makes an organism what it is. And we marveled at how simple the answer seemed to be. But now, in the call for a functional genomics, we can read at least a tacit acknowledgment of how large the gap between genetic "information" and biological meaning really is.

Of course, the existence of such a gap had long been intuited, and not infrequently voices could be heard attempting to caution us. It is only now, however, that we begin to fathom its depths, marveling not at the simplicity of life's secrets but at their complexity. One might say that structural genomics has given us the insight we needed to confront our own hubris, insight that could illuminate the limits of the vision with which we began.

In the main body of this book, I review four of the more important lessons that molecular genomics has helped us learn. The first concerns the role of the gene in what may well be the most fundamental dynamic of the living world: maintaining the faithful reproduction of traits from generation to generation and providing the variability on which evolution depends—that is, ensuring both genetic stability and genetic variability. In the second chapter, I discuss the meaning of gene function and ask: What is it that a gene *does?* In the third, I examine the notion of a genetic program and contrast that idea with the concept of a developmental program. And in the fourth chapter, I argue for the importance of resiliency in biological development and consider the ways in which a search for design principles that would ensure developmental reliability and robustness exposes some of the limits of genetic analysis.

Throughout each of these chapters, my primary focus is on the ever-widening gaps between our starting assumptions and the actual data that the new molecular tools are now making available. These tools are themselves the direct product of the most recent advances in molecular genetics

and genomics; yet at the same time, and in the most eloquent testimony to the prowess of science I can imagine,
they have worked to erode many of the core assumptions on
which these efforts were first premised. In the recent calls
for a functional genomics, I read an acknowledgment of
the limitations of the most extreme forms of reductionism
that had earlier held sway. And even though the message has
yet to reach the popular press, to an increasingly large number of workers at the forefront of contemporary research, it
seems evident that the primacy of the gene as the core explanatory concept of biological structure and function is
more a feature of the twentieth century than it will be of the
twenty-first. What will take its place? Indeed, we might ask,
will biology ever again be able to offer an explanatory framework of comparable simplicity and allure?

What, in short, will the biology of the twenty-first century look like? I have no crystal ball, but perhaps some indications of its shape can be seen in the new lexicon that begins to emerge as biologists turn their attention to "cross-
talk" and "checkpoints," to genetic, epigenetic, and "post-
genomic" metabolic networks, and even to multiple systems
of inheritance. But will the new lexicon ever cohere into an
explanatory framework providing anything close to the satisfaction that genes once offered? This I cannot say, and in
any case, the answer will depend not only on what biologists
find, not only on the adequacy of such terms and concepts
to these findings, but also on the particular needs those explanations will be expected to satisfy in the coming decades.

Only three predictions seem safe to make about the
character of biology in a post-genomic age. First, a radically
transformed intra- and intercellular bestiary will require ac-
commodation in the new order of things, and it will in-

clude numerous elements defying classification in the traditional categories of animate and inanimate. Second, biologists who seek to make sense of these new elements will have a considerably expanded array of conceptual tools with which to work. Third, even so, they are not likely to stop talking about genes—not, at least, in the near future.

Why is that? What is it that keeps the term alive? This question I take up in my conclusion, and, in brief, my answer is twofold. First, Johannsen's "little word" has become far too entrenched in our vocabulary for it to disappear altogether; and second, despite all its ambiguity, it has not yet outlived its usefulness. Thus, at the end of this book I turn to the question "What are genes *for?*" and argue that to ask this question is also, at least implicitly, to ask "What is gene *talk* for?" I point to several particularly important ways in which gene talk functions today.

Paramount among these is the convenience of gene talk as an operational shorthand for scientists working in specific experimental contexts. Furthermore, gene talk identifies concrete levers or handles for effecting specific kinds of change. And finally, gene talk is an undeniably powerful tool of persuasion, useful not only in promoting research agendas and securing funding but also (perhaps especially) in marketing the products of a rapidly expanding biotech industry. My rather brief comments about these functions are not intended as a recapitulation of the central arguments of the book but rather as a way of calling attention to some of the many questions and issues it does not address, and for which the interested reader will need to look elsewhere.

## 1

## Motors of Stasis and Change:
## The Regulation of Genetic Stability

How can we, from the point of view of statistical physics, reconcile the facts that the gene structure seems to involve only a
comparatively small number of atoms . . . and that nevertheless
it displays a most regular and lawful activity—with a durability
or permanence that borders upon the miraculous?
  Let me throw the truly amazing situation into relief once
again. Several members of the Hapsburg dynasty have a peculiar disfigurement of the lower lip ("Hapsburger Lippe") . . .
Fixing our attention on the portraits of a member of the family
in the sixteenth century and of his descendant, living in the
nineteenth, we may safely assume that the material gene structure responsible for the abnormal feature has been carried on
from generation to generation through the centuries, faithfully
reproduced at every one of the not very numerous cell divisions
that lie between . . . How are we to understand that it has remained unperturbed by the disordering tendency of the heat
motion for centuries?

ERWIN SCHROEDINGER, *What Is Life?* (1944)


If the Mendelian revolution marked the
turning point of twentieth-century biology, then surely the
Darwinian revolution was the great watershed of the nine-

teenth century. The realm of living organisms could no longer to be fitted into a great "Chain of Being"; it required its own figuration: more of a tree than a chain, and as much a succession of becoming as of beings. The living world became a world in time, and both its occupants and its relational structure were reconfigured as products of its evolutionary history. After the publication of *On the Origin of Species* in 1859, few could be found among the scientifically literate who still believed in the fixity of species. Moreover, Darwin's evolutionary theory offered his readers a mechanism for the origin and transformation of species—natural selection acting upon individual variation. Yet, for all the power of that theory, a fundamental mystery remained. If change is the essence of life, how are we to account for the remarkable stability with which, in each generation, organisms develop and grow true to the type of their particular species, and with a certainty that endures over the lifetime of that species?

Viewed from the perspective of geological time, species transform and evolve. Yet viewed from the perspective of historical time, they display an unmistakable constancy in form and function. But on this matter—on the "stability of type" (to borrow a phrase from Francis Galton) that is so conspicuously maintained over the course of generations— Darwin's theory was silent. However eloquently and powerfully the theory of evolution by means of natural selection might account for changes in biological form and function occurring over eons and reflected in the geological record, it could not begin to explain the reproducibility of that same form and function over the shorter spans of genealogical time. Nor could it offer any account of the persistence of particular individual features from generation to generation, of the clearly recognizable family resemblances that are passed on from parents to offspring.

Of course, Darwin was not privy to the insights of genetics, nor could he have been. He shared with his contemporaries a belief in "blending heredity"—the view that the characteristics of an offspring are, somehow, a blend of the parents' characteristics—but he had nothing to say about how such distinctive features as the Hapsburg lip might endure without dilution. Nor could he offer any kind of answer to the dilemma that was later to plague Schroedinger: How can we understand the reproduction of individual features, generation after generation, with such fidelity as to lend them a "durability or permanence that borders upon the miraculous?"

The fact is that Darwin's preoccupations were different. Throughout his life, he focused his attention on mechanisms of transformation; the mechanisms required for conservation eluded both his understanding and, for the most part, his interest. And while he acknowledged that "our ignorance of the laws of variation is profound" and devoted considerable attention to the ways in which the variation essential to natural selection might arise, nowhere did he express concern about a corresponding ignorance of the laws of constancy.[1]

The task of searching for the laws of constancy—that is, of accounting for intergenerational stability—thus fell to Darwin's heirs. Indeed, the century of the gene begins with this task—or more specifically with efforts to account for the persistence of individual traits through the generations. Of course, just as with any collective endeavor, the science of genetics arose out of multiple needs and a variety of different interests, and these have been well chronicled by many historians. My focus here, in Chapter 1, is on the particular force that the search for constancy of individual traits exerted on the origins of the very concept of the gene. A crucial component of that concept, I argue, enters the history of genetics even before the word *gene* was coined, and it enters with the supposition that underlying each individual trait is a hereditary unit so stable that its stability can account for the reliability with which such traits are transmitted through the generations. In other words, the problem of trait stability was answered by assuming the existence of an inherently stable, potentially immortal, unit that could be transferred intact through the generations.

In the first part of this chapter, I trace the increasing hold this assumption of the intrinsic stability of hereditary elements came to have on geneticists in the first part of the century, its apparent vindication in the middle of the century, and its gradual dissolution over the last few decades. To be sure, genetic stability remains as remarkable a property as ever, and it is clearly a property of all known organisms. The difficulty arises with the question of how that stability is maintained, and this has proven to be a far more complex matter than we could ever have imagined. Furthermore, we will see that the maintenance of genetic stability turns out to be inextricably bound up with the generation of variability. Thus, in the second part of this chapter, I return to Darwin's concerns, taking up the companion issue of transformation and discussing some of the surprising challenges that new research on mechanisms of conservation pose to the simple neo-Darwinian picture of evolution by the cumulative operation of natural selection on randomly generated small mutations.

Finally, a word about the relation between the stability

of "type" (that is, the stability with which organisms, in each generation, develop and grow true to the type of their particular species) and the stability of individual traits. For a long time, it was assumed that genes are as capable of explaining the development of individual traits as they are of explaining the development of whole organisms, and therefore that genetic stability sufficed to account for what I will later on in this book call *developmental stability*. I use the term to refer to the reliability with which organisms of a particular species undergo the passage from fertilization to maturity, generation after generation, each time reproducing a phenotype that is clearly recognizable as characteristic of that "type." Thus, while genetic stability is a property of all organisms, developmental stability is a term primarily applicable to multicellular organisms that pass through embryonic stages of development—that is, metazoan organisms. The differences between these two kinds of stability may be significant, but discussion of such differences must be deferred until after I have said more about the relation between genes and development. Accordingly, in my fourth and final chapter I return to the particular challenges raised in attempting to account for developmental stability.

### Explaining Genetic Stability

August Weismann (1834-1914)—one of the great zoologists of the latter part of the nineteenth century—put the problem succinctly: "When we find in all species of plants and animals a thousand characteristic peculiarities of structure continued unchanged, through long series of generations; when we even see them in many cases unchanged throughout whole geological periods; we very naturally ask for the causes of such a striking phenomenon ... How is it that ... a single cell can reproduce the tout ensemble of the parent with all the faithfulness of a portrait?"[2] In these brief remarks, written in 1885, Weismann defined the challenge for a science of heredity—indeed, one might read the entire history of genetics as an attempt to answer the question he posed. But Weismann did more than pose the question: he also proposed something of an answer, and the form of his answer helped set the science of heredity on the particular track it would follow for the next sixty years or more.

Whatever the mechanism by which a single cell reproduces the traits of the parent, Weismann assumed the existence of particulate, self-reproducing elements that "determine" the properties of an organism; appropriately enough, he called these elements determinants. This assumption was hardly unique to Weismann—in fact, Darwin himself had hypothesized the existence of some such elements (his gemmules). The Dutch botanist Hugo de Vries, a near-contemporary of Weismann's (1848–1935), also hypothesized the existence of elementary hereditary units. As he wrote, "Just as physics and chemistry are based on molecules and atoms, even so the biological sciences must penetrate to these units in order to explain by their combinations the phenomena of the living world."[3] De Vries called his units pangens, a term he introduced in 1889 in an effort to salvage the best of both Darwin's gemmules and Weissman's determinants.

But Weismann assumed more than the existence of elementary hereditary units. In order to explain the remarkable fidelity with which such traits were reproduced generation after generation, he further hypothesized the sequestration of a full complement of these elements in a substance "of a definite chemical, and above all, *molecular composition*." He called this substance the "germ-plasm" and argued that a germ-plasm, insulated from the ravages of individual mortality, could be transferred, intact, from one generation to another. Thus he wrote, "I have attempted to explain heredity by supposing that in each ontogeny, a part of the specific germ-plasm contained in the parent egg-cell is not used up in the construction of the off-spring, but is reserved unchanged for the formation of the germ-cells of the following generation."[4] Weismann's theory traveled wide and fast. In his influential textbook published only a few years after Weismann's work had appeared in English, the American zoologist E. B. Wilson wrote, "As far as inheritance is concerned, the body is merely the carrier of the germ-cells, which are held in trust for coming generations."[5]

Experimental biology was still in its infancy at the end of the nineteenth century, and Weismann had no way of knowing what these hereditary elements might be. Nor did de Vries, or any other student of heredity at that time. This was a period of grand speculations, and Weismann's were among the grandest. As he explained his philosophy, "Biology is not obliged to wait until Physics and Chemistry are completely finished; nor have we to wait for the investigation of the phenomena of heredity until the physiology of the cell is complete . . . Science is impossible without hypotheses and theories; they are the plummets with which we test the depth of the ocean of unknown phenomena and thus determine the future course to be pursued on our voyage of discovery."[6] Given how little they had to go on in the way of concrete evidence, it comes as small surprise to find how much (or how sharply) these early thinkers about heredity differed from one another both in their characterization of hereditary elements and in their conjectures about how these elements could impress their various characteristics on the formation of particular cells and tissues. What is more surprising is how much they shared. Underlying all their differences were two enduring articles of faith.

The first of these was that, just as atoms and molecules provided the fundamental units of explanation in physics and chemistry, so too would particulate hereditary elements serve as the fundamental units of biological explanation. These units might themselves be some kind of atom or molecule, or they might be made up of molecules, but the important point was that they were elemental, the primitive units with which the study of heredity must begin.

The second article of faith was closely related, and it held that responsibility for intergenerational stability inhered in the fixity of these material elements, taken either as individual units or in their collective composition. For Weismann, the burden of stability lay in the sequestration of a certain substance "of a definite . . . molecular composition" in a protected lineage of germ cells, where they would be held inviolate for future generations. For de Vries, it lay in the sequestration of the individual particles in the nucleus of each and every cell, with one particle representing one hereditary characteristic. But once sequestered, whether in the germ-plasm or in the nucleus, the fixity of the elements themselves was simply taken for granted, accepted as part of their definition.

The rediscovery of Mendel's rules of inheritance in 1900

marked the beginning of an end to the era of grand speculation in the study of heredity. Indeed, Johannsen's aim in
coining the term *gene* in 1909 was to mark a break with the
preconceptions of his predecessors. "The word 'gene,' he
wrote, "is completely free from any hypotheses."[7] But it
takes more than a new word to effect a complete break with
the past. Weismann's determinants and de Vries' pangens
were still the direct precursors of the gene, and inevitably
some of the preconceptions underlying these earlier concepts carried over. Genes were hypothetical entities, but, like
their precursors, they were particulate entities (Mendel himself had called his factors *Elemente*). Furthermore, whatever
they were made of—indeed, even for those who thought of
them as no more than a bookkeeping device—the capacity
for faithful transmission from generation to generation remained built into the very notion, as it were, by definition.

No student of heredity, either before or after the watershed of 1900, thought of these hereditary elements literally
as atoms, but the analogy with the fundamental units of
physics and chemistry continued to lurk in people's minds.
As E. B. Wilson wrote in 1923, "Even if considered only as
working instruments . . . these conceptions have a practical
value almost comparable to that of the atomic theory as employed in chemistry and physics."[8] To the extent that the
very notion of an atom implied stability, the analogy would
have seemed especially apt for thinking about the immutability of hereditary elements. But even after 1901, when it
had first been observed that the elements of physics and
chemistry could themselves undergo spontaneous "transmutation," physicochemical elements continued to serve as
models for the elements of heredity—perhaps, with this new

possibility, as even better models. In fact, the occasional spontaneous transmutation of atoms that were nevertheless essentially stable served biologists well, for it opened a way to reconcile genetics with evolution. Hereditary elements too must sometimes change—indeed, it was precisely the occasional occurrence of such changes (or mutations) that made experimental genetics possible in the first place, for it was the tracking of mutations through generations that constituted the core method of classical genetics. In this sense, the advent of quantum mechanics might be said to have been fortuitous for biologists, especially for those who continued to look to physics as a model for their own fundamental units.

The geneticist H. J. Muller was one. In 1921 he wrote: "It is not physics alone which has its quantum theory. Biological evolution too has its quanta—these are the individual mutations."[9] Five years earlier, while still a student, Muller had already noted "the curious similarity which exists between the main problems of physics and of biology." Furthermore, he argued, finding the means to influence mutation "might obviously place the process of evolution in our hands," just as the power to direct the transmutation of the elements could render "inanimate matter practically at our disposal." In conclusion, he proclaimed, "Mutation and Transmutation—the two keystones of our rainbow bridges to power!"[10]

Muller's contributions to the history of experimental genetics are legendary, but he was also a theoretician and a visionary. And in neither his theoretical nor his visionary writings did he ever lose sight of what remained, for him, the central question: Just what sort of entity is a gene? Nor was he ever able to provide an answer.

## SCHROEDINGER'S QUESTION

The rise of classical genetics over the first half of this century is one of the great success stories of our time, and its history has been well documented. Yet despite its many successes, the question remained: What kind of object might a gene be that it can reproduce itself with such remarkable fidelity, generation after generation? Indeed, it was this very property of the gene, its manifestation of "a durability or permanence that borders upon the miraculous" that so mystified the physicist Erwin Schroedinger in the early 1940s as to inspire him to take on that grandest of all questions, "What is life?"[11] To Schroedinger, it seemed evident that the question of what endowed the gene with such durability, what lent it its apparent immunity to the second law of thermodynamics—with a "permanence unexplainable by classical physics"—got at the very core of the distinction between living and nonliving beings. He believed not only that the answer to this question would solve the problem of heredity but also that it would explain the equally remarkable capacity of organisms to maintain themselves against the ravages of entropy, to keep on going for so much longer than the laws of physics would lead us to expect. It would give us, in short, the secret of life.

Schroedinger, alas, did not find the secret of life. As one of the fathers of quantum mechanics, he not surprisingly sought the solution of this problem in the explanation that theory had already provided for the chemical stability of molecules. The particular model of gene structure on which he based his hope had been proposed in 1935 by two physicists and a geneticist.[12] In their picture, the gene was figured as a quantum mechanical system that derived its stability from the height of the energy barrier separating one state from another. The theoretical contributions to the model were made by Max Delbrück (a student of Niels Bohr), and accordingly Schroedinger referred to it as "Delbrück's model," adding to Delbrück's speculations the provocative proposal that the gene is not just a large molecule but an "aperiodic crystal or solid." Indeed, he saw "no alternative to the molecular explanation of the heredity substance." As he wrote, "The physical aspect leaves no other possibility to account for its permanence. If the Delbrück picture should fail, we should have to give up further attempts."[13]

Yet the Delbrück picture did fail, and with that failure so too did Schroedinger's solution. Nonetheless, even with all its defects, the very effort of so prominent a physicist to solve so fundamental a biological problem served as powerful inspiration for an entire generation of young physicists and biologists, encouraging them in their own efforts to find the molecular structure of the gene. And soon they succeeded. Success, however, came not as a consequence of theoretical speculation but out of a series of experimental reports that narrowed the search to the structure of a specific chemical candidate.

The route by which biologists came to accept DNA as the genetic material has a long, rich, and well-documented history.[14] In most popular accounts, however, that history begins with the paper by Avery, MacLeod, and McCarty which demonstrated through direct experiment that DNA was the carrier of biological specificity (at least in bacteria).[15] This now-classic paper was published in the same year as Schroedinger's book. In it, the authors provided strong evidence arguing that DNA "must be regarded not merely as structurally important but as functionally active in determining the biochemical activities and specific characteristics of pneumococcal cells."[16] But not everyone was immediately persuaded. Indeed, it was only after the almost equally famous "blender experiment" of Hershey and Chase in 1952 that most biologists were won over to the view that the genetic material was made up of DNA.[17]

Less than a year later, Watson and Crick struck gold. When the last piece of their model for the structure of DNA fell into place in the spring of 1953, Watson tells us that Crick "winged into the Eagle to tell everyone within hearing distance that we had found the secret of life."[18] It is not hard to understand his enthusiasm. Not only did that structure provide a mechanism for the gene's remarkable capacity for self-replication—a mechanism that was stunning in its very simplicity—but also, and at the same time, it provided an (equally simple) explanation for the stability of the gene—for the ostensibly miraculous fidelity with which it could be copied over so many generations. Complementary base-pairing could, at one fell swoop, do the work of both replication and conservation, or so it seemed (Figure 1).

If one assumed that DNA was an intrinsically stable molecule (as people did) and that complementary base-pairing proceeded without error, then nothing more would be required. In a sense, one might even say that Watson and Crick's triumph provided retrospective vindication of Schroedinger's own earlier speculations. From the vantage point of the simple picture that now emerged, his proposal of an aperiodic crystal or solid for the structure of the gene (and perhaps of the entire chromosome) acquired, at least in hindsight, an aura of prophecy.

Watson and Crick's achievement stands unrivaled in the annals of twentieth-century biology, and it is worth pausing

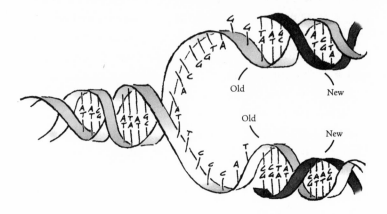

Figure 1: DNA replicating itself. A simplified representation of semiconservative replication of DNA, in which each strand of the original molecule acts as a template for the synthesis of a new complementary DNA molecule, following the rules of complementary base pairing: adenine (A) to thymine (T), and guanine (G) to cytosine (C). Two strands of DNA are thus obtained from one, identical to one another and to the parent molecule. *(By Nick Thorkelson.)*

for a moment to register the extraordinary sense of satisfaction that accompanied their findings. Since the beginning of the century, the notion of the gene as a self-replicating entity that carried the secret of its immortality in its very structure had been a staple of genetics, but no one had ever been able to say what kind of material such an entity might be made of. Now, after more than fifty years, an actual chemical substance—one already known to be a basic constituent of chromosomes—had been shown to have the necessary defining properties. Even before a mechanism was worked out by which the sequence of nucleotides in a DNA molecule could be translated into a sequence of amino acids in a protein molecule, confidence was widespread that the material basis of genetics had finally been established.

The decade that followed seemed little short of heroic. All the fundamental problems of biology yielded quickly, without difficulty or surprise. In 1968 an article appeared in *Science* entitled "That Was the Molecular Biology That Was." Here, Gunther Stent, an active participant in the exciting new research, described the approaching decline of the discipline that was "only yesterday an avant-garde but today definitely a workaday field."[19] In Stent's view, by 1963 molecular biology had already entered what he called its "academic phase." He wrote, "All hope that paradoxes would still turn up in the study of heredity had been abandoned long ago, and what remained now was the need to iron out the details."[20]

### Ironing Out the Details

The history of science is replete with irony, and the aftermath of Watson and Crick's tour de force offers no exception. As everyone now knows, Stent could not have been more wrong. Molecular biology's course after 1968 was anything but a decline. Only two years later, with the isolation of a restriction enzyme that can recognize and cut DNA molecules at specific sites, the field was launched into a new, and in some ways even more productive, era. Restriction enzymes are the basis of the powerful techniques of recombinant DNA that have opened vast new vistas and, in doing so, have yielded so many surprises.

One such surprise bears directly on Weismann's original question, that is, on the source of genetic stability. To be

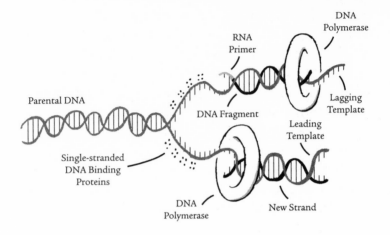

**Figure 2: The collaboration of proteins during replication.** The replication of DNA requires the collaboration of many different proteins: (1) The two parent strands are unwound with the help of DNA helicases. (2) Single-stranded DNA binding proteins attach to the unwound strands, preventing them from winding back together. (3) The strands are held in position, binding easily to DNA polymerase, which catalyzes the elongation of the leading and lagging strands. (4) While the DNA polymerase on the leading strand can operate in a continuous fashion, RNA primer is needed repeatedly on the lagging strand to facilitate synthesis. DNA primase, which is one of several polypeptides bound together in a group called primosomes, helps to build the primer. (5) Each new fragment is attached to the completed portion of the lagging strand in a reaction catalyzed by DNA ligase. *(By Nick Thorkelson.)*

sure, DNA is copied in living cells with a fidelity that borders on the miraculous. But contrary to expectations, the structure of DNA provides only the beginning of an explanation for this high fidelity. In fact, left to its own devices, DNA cannot even copy itself: DNA replication will simply not proceed in the absence of the enzymes required to carry out the process (Figure 2). Moreover, DNA is not intrinsically stable: its integrity is maintained by a panoply of proteins involved in forestalling or repairing copying mistakes, spontaneous breakage, and other kinds of damage incurred in the process of replication. Without this elaborate system of monitoring, proofreading, and repair, replication might proceed, but it would proceed sloppily, accumulating far too many errors to be consistent with the observed stability of hereditary phenomena—current estimates are that one out of every hundred bases would be copied erroneously. With the help of this repair system, however, the frequency of mistakes is reduced to roughly one in 10 billion (Figure 3).[21]

In point of historical fact, however, indications that the cell was involved in the maintenance of genetic stability had begun to emerge well before the "recombinant DNA revolution," even if they failed to attract much attention. The first signs came in the late 1950s and early 1960s, from studies of radiation damage in bacteria and bacterial viruses (phages) at Oak Ridge National Laboratory, and especially from the discovery that certain kinds of damage could be spontaneously reversed. Bernard Strauss, who was a participant in this early work, writes, "The discovery that genic material did not stand permanently aloof from cellular metabolism was a major surprise of the 1950's and 1960's."[22] Yet in the wider community of molecular biologists, the surprise of these new findings was slow to register and their implications were even slower to dawn.

Rollin Hotchkiss of the Rockefeller Institute was a prominent member of that wider community, but he was also something of an exception. As early as 1968, he wrote: "We are turning away from the DNA of a decade just over, a relatively unchanging, stable reservoir of linear information. It has had its convincing tellings and smugly one-dimen-

Figure 3: **Repair mechanisms.** Cartoon depiction of the basic mechanisms of nucleotide selection, proofreading, and mismatch and excision repair involved in ensuring the fidelity of replication. *(By Nick Thorkelson.)*

sional retellings and become 'well known' (which is to say, often mentioned). But it has become necessary to face the fact that DNA grows, issues directives, opens up, closes, twists, and untwists. We are coming to realize how marvelously communicative it is, and that it is not an aloof, metabolically inert material, but instead one maintained and exquisitely balanced in an actively supported status quo."[23]

A more typical response, however, is suggested by Franklin Stahl, one of the central figures in the heroic age of molecular biology. In a recent history of the discovery of DNA repair mechanisms, Errol Friedberg (another participant in the early repair field) reports Stahl's response when

Friedberg "challenged him with the question of why the concept of gene/DNA repair was late in coming": "I suspect because of a widespread belief (unspoken I suspect, but amounting to worship) among geneticists that *the genes* are so precious that they must *(somehow)* be protected from biochemical insult, perhaps by being carefully wrapped. The possibility that *the genes* were dynamically stable, subject to the hurly-burly of both insult and clumsy (i.e., enzymatic) efforts to reverse the insults, was unthinkable."[24]

When I later queried him further, Stahl acknowledged that much of the early work on repair in radiation biology was overlooked.[25] But to the widespread belief that stability inhered in the gene itself he added two other factors: one disciplinary ("the investigators usually seemed to have little sense of genetics") and the other political (radiation biology was "somewhat suspect" because it was under the sponsorship of the Atomic Energy Commission). For Stahl, it was only after Evelyn Witkin and Miroslav Radman's work implicating recombination with repair (discussed below) that "we could no longer blow off investigations on repair." "Now, not only was the geneticists' domain of recombination invaded by the repair people but their domain of mutagenesis was also. The walls were not only breached, but they were toppled."[26]

Over the last fifteen years the field of DNA repair has truly exploded. In 1994 the journal *Science* gave its "Molecule of the Year" award to the enzymatic repair machinery, and for good reason. The mechanisms already known to be involved in proofreading, editing, and repairing damaged or miscopied DNA can scarcely fail to astonish us—by their ingenuity, their complexity, and perhaps especially by their implications for our understanding of evolution. But they are still far from clearly understood, and, as is inevitably the case with cutting-edge research, as yet subject to considerable debate. For all these reasons, the sketch that follows is simultaneously technical and provisional, and in both cases unavoidably so.

Three different kinds of processes seem to be involved in ensuring the fidelity of replication as it proceeds. The first works by helping to select the correct nucleotide for complementary binding. The second works by checking the most recently added nucleotide and immediately removing it if it should fail the test of complementarity. The third comes into action only after a new strand has been synthesized, and it works by repairing mismatches that might have occurred in spite of the first two error-avoidance mechanisms. A fourth set of repair mechanisms—first observed in the early work on photo-reactivation—come into play later, in response to environmentally inflicted damage (caused, for example, by ultraviolet light). If the damage has been confined to a single strand, these excision repair mechanisms can reverse it with little chance of error by excising the damaged section and allowing it to be recopied from the undamaged strand (Figure 3).[27]

The stability of gene structure thus appears not as a starting point but as an end-product—as the result of a highly orchestrated dynamic process requiring the participation of a large number of enzymes organized into complex metabolic networks that regulate and ensure both the stability of the DNA molecule and its fidelity in replication.[28] As the late Robert Haynes has written, "The stability of genes is now seen to be more a matter of biochemical dynamics, than of the molecular 'statics' of DNA structure. The genetic machinery of the cell provides the most striking example known of a highly reliable, dynamic system built from vulnerable and unreliable parts."[29]

## THE LIMITS OF GENETIC STABILITY

Even with such an elaborate process of proofreading and repair, genetic stability is not absolute, and fortunately not. If genes were truly immortal, and if their replication proceeded with perfect fidelity, the evolution of new genetic structures would never have been possible. As Darwin so clearly understood, change too is a desideratum of life, and the question naturally arises: How much genetic instability (or mutability) is necessary? How much would be required to accord with the pace at which evolution has actually occurred? Indeed, is it possible that the balance between genetic stability and mutability observed in the organisms we see today is itself a product of evolution? In other words, might selective pressures have operated on the very capacity to evolve, giving rise to the evolution of special mechanisms for generating change?

Evolution by natural selection depends on the occurrence of that rarest of events, mistakes that proved beneficial. But the fact remains that the vast majority of naturally occurring mistakes are either harmful or neutral.[30] Would it not therefore have been advantageous to the survival of cells and organisms to have developed mechanisms ensuring an even higher accuracy of replication, permitting fewer mistakes?

Given what is known, it is not hard to imagine ways of increasing genetic stability. But one of the most interesting insights to come out of work on repair mechanisms is the recognition that the advantage of increased fidelity in replication is not fixed but rather depends on both the organism and the conditions in which that organism finds itself. The critical dependence of genetic stability on proofreading and repair enzymes may have come as a great surprise, but more surprising yet was the discovery of "repair" mechanisms that sacrifice fidelity in order to ensure the continuation of the replication process itself—and hence the survival of the cell. Far from reducing error, such mechanisms actively generate variations in nucleotide sequence; moreover, it appears that when and where they come into play is itself under genetic regulation. With such findings as these,

Barbara McClintock's remarks in her 1983 Nobel address describing the genome as "a highly sensitive organ of the cell, monitoring genomic activities and correcting common errors, sensing the unusual and unexpected events, and responding to them" no longer seem quite so far-fetched.[31]

The first indications of an error-prone mechanism of repair also came out of early studies of UV radiation damage in bacteria, and once again the implications were slow to dawn. Evelyn Witkin was a key figure in this work. In 1967, after isolating a mutant of *E. coli* in which this activity was repressed, she argued for a mechanism of stress-induced mutagenesis that might itself be under genetic control. Soon after at the Massachusetts Institute of Technology, Maurice Fox conducted a preliminary experiment showing that a closely related error-prone repair activity implicated in the reactivation of UV-irradiated bacterial viruses (and known as Weigle-reactivation) depends on the synthesis of new protein—an observation that led him to suggest to the young Miroslav Radman that one or more specific genes might be involved in the induction of error-prone repair.[32] Four years later, Radman coined the expression "SOS response" to designate both this and Witkin's UV-induced error-prone (or hypermutagenic) modes of replication.[33]

For Radman, son of a fisherman, the meaning of SOS was clear: "An international distress signal to save endangered life on the sea."[34] The term stuck and is now widely understood as referring to last-ditch efforts invoked to save the cell from going under. In his first publication on the subject, however, Radman was somewhat more circumspect: "Because of its 'response' to DNA-damaging treatments," he wrote, "we call this hypothetical repair 'SOS repair' . . . In order for SOS repair to function it should require specific genetic elements, the inducing signal and *de novo* protein synthesis."[35] The questions that needed answering were obvious. First, what are these specific genes and proteins? And, second, how are they induced? But without the tools with which to answer such questions, further progress proved desultory, and soon even Radman's attention turned to other problems.

A quarter of a century later, the regulation of mutability has become one of the hottest topics in molecular biology; and with the new analytic techniques that have now become available, many aspects of the biochemical machinery involved in such regulation have been elucidated. But with every step toward elucidation, the picture is rendered ever more complex by the increasing wealth of detail. Scores of proteins have already been implicated in the cell's response mechanisms, and reports of new players appear with every passing month.

Furthermore, as the picture becomes more and more complicated, so too does it appear ever more radical in its implications. To be sure, there is nothing radical in the notion that genetic stability and variability (or mutability) are complementary in their effects. Now, however, stability and mutability are proving to be flip sides of one another in the specific mechanisms by which they are controlled. Both are at the mercy of enzymatic processes, and apparently equally so. Moreover, not only are the mechanisms controlling stability and mutability held in a delicate balance, but that very balance is under cellular regulation, and it shifts in response to the particular environment in which the cell finds itself.

All this is a far cry from the traditional view of DNA as

an inherently stable molecule subject to occasional random errors, and it suggests an even further departure from the traditional view of evolution as a process of cumulative selection of those exceedingly rare mutations that happen to result in increased fitness. At least to many, the new picture seems to accord far better with McClintock's image of the genome as "a highly sensitive organ" than it does with the neo-Darwinian view of the genome as a strictly passive partner in the evolutionary two-step of variation and natural selection.

That regulation of genetic stability and mutability is a feature of all living systems is now widely accepted. The phenomenon has been most intensively studied in bacteria, and here a number of mechanisms that enhance mutation rates are at least partially understood. In many cases, increased mutation rate is associated with a defect in genes required for the overall maintenance of genetic stability. Such genes are sometimes called *mutator genes,* although the terminology is a bit misleading because mutation rates increase only when the gene is defective. For example, a mutation may eliminate or reduce the efficacy of an enzyme required for proofreading or error correction; it may interfere with the process by which newly synthesized strands of DNA can be distinguished from the old strands (for example, by methylation of the old strand); or it may eliminate or disable an enzyme involved in excision repair. Alternatively, a mutation might lead to increased mutation rates by disabling the regulatory mechanism that represses the activity of the error-producing SOS system under normal conditions—that is, in the absence of extensive (stress-induced) damage.

It is important to note, however, that the unmutated (or wild-type) SOS system becomes a generator of increased mutation rates when called into action by conditions of severe stress.[36] In cultures of bacteria that have not undergone deregulating mutations (that is, in wild-type cultures), the SOS system is activated only by specific signals (generated, for example, by the persistent presence of single-stranded DNA or interruptions in the process of replication). Such signals activate a variety of mechanisms that allow the damaged region to be by-passed, filled in, or exchanged with a homologous region of DNA nearby (through recombination). Simultaneously, they lead to the inactivation of some of the normal proofreading functions, thus permitting replication to proceed even with the accumulation of many mistakes. Once replication has been completed, however, the SOS system returns to its normal repressed state and the machinery of proofreading and error-correction resumes its customary vigilance.

Most of these SOS functions are directly implicated in repair of one sort or another—in the sense, that is, that they make it possible for replication to proceed. But Radman and his colleagues now argue that the SOS system also functions merely to generate diversity, as if for its own sake, without serving any obvious repair function.[37] Perhaps he is right, but how could one possibly make evolutionary sense of such an idea?

## The Evolution of Evolvability—Molecular Biology's Challenge to Neo-Darwinism

We now know that mechanisms for ensuring genetic stability are a product of evolution. Yet a surprising number of mutations in which at least some of these mechanisms are disabled have been found in bacteria living under natural conditions. Why do these mutants persist? Is it possible that they provide some selective advantage to the population as a whole? Might the persistence of some mutator genes in a population enhance the adaptability of that population? Apparently so. New mathematical models of bacterial populations in variable environments confirm that, under such conditions, selection favors the fixation of some mutator alleles and, furthermore, that their presence accelerates the pace of evolution.[38] Recent laboratory studies of bacterial evolution provide further confirmation,[39] lending support to the notion that organisms have evolved mechanisms for their own "evolvability."[40]

Mutator genes, however, are constitutive—that is, they give rise to high mutation rates even in the absence of provocation. Thus, over time they might enhance the adaptability of a population, but unlike inducible systems such as the SOS system, they offer no obvious adaptative advantages to individual organisms.[41] Might there be a connection between these quite different mechanisms for generating rapid change? Radman believes there is. He writes: "Mutagenesis has traditionally been viewed as an unavoidable consequence of imperfections in the process of DNA replication and repair. But if diversity is essential to survival, and if mutagenesis is required to generate such diversity, perhaps mutagenesis has been positively selected for throughout evolution."[42] In support of this view, he cites the recent identification of several enzymes involved in the SOS system. Such mutases, as Radman calls them, are "designed to generate mutations."[43] And because they are inducible, they can be argued to enhance the resources of the organism as well as those of the population for coping with unanticipated environmental challenges. "Chance," as one of the organizers of a recent conference on "Molecular Strategies in Biological Evolution" puts it, "favors the prepared genome."[44]

The notion that mechanisms for evolvability could themselves have evolved is a serious provocation for neo-Darwinian theory, for it carries the heretical implication that organisms provide not just the passive substrate of evolution but their own motors of change; it suggests that they have become equipped with a kind of agency in their own evolution. It also strongly implies the operation of selection on levels higher than the gene, and higher even than the individual organism. As James Shapiro writes, "These molecular insights lead to new concepts of how genomes are organized and reorganized, opening a range of possibilities for thinking about evolution. Rather than being restricted to contemplating a slow process depending on random (i.e., blind) genetic variation and gradual phenotypic change, we are now free to think in realistic molecular ways about rapid genome restructuring guided by biological feedback networks."[45]

Shapiro, like Radman, studies bacteria. But the idea that organisms have evolved their own mechanisms for change has come to extend far beyond bacteria, and it has been gathering increasing support from biologists across a wide range of specialties. As far as I can tell, the first use of the expression "the evolution of evolvability" was as the title of a paper on "artificial life" by Richard Dawkins.[46] Dawkins wrote, "A title like 'The Evolution of Evolvability' ought to be anathema to a dyed-in-the-wool, radical neo-Darwinian like me! Part of the reason it isn't is that I really have been led to think differently as a result of creating, and using, computer models of artificial life."[47] It was by tinkering with such computer simulations that Dawkins was led to hypothesize "a kind of higher-level selection, a selection not for survivability but for evolvability."[48] For many others, however, what led them to the notion was trying to make sense of their own and their colleagues' experimental observations. This is especially the case for the increasing number of developmental and evolutionary biologists seeking to bring some coherence to the accumulating mass of data attesting to both conservation and transformation of cellular and developmental systems across the evolutionary spectrum.

As the term is employed in these wider communities, *evolvability* refers to the capacity to generate any kind of heritable phenotypic variation upon which selection can act. It may be based on individual mutator genes or on higher-level genetic or epigenetic networks. The particular appeal of such an extended sense of the notion of evolvability lies in its ability to shed light on the evolution of developmental systems; Per Alberch may have been the first to make this point.[49] But probably the most extensive argument made to date for the role of evolvability in the evolution of complex organisms comes from John Gerhart and Marc Kirschner. As they wrote in their 1997 book, *Cells, Embryos, and Evolution,* "Throughout [the history of genetics], the organism remained a black box, translating random change in its genes into phenotypic variation to be acted on by selection. This black box is being quickly opened up by modern biology. In it we find that the connections between genotype and phenotype have been crafted by evolution to collaborate with evolution."[50]

The particular concern of these authors is with the explosion of diversity in the evolution of metazoan organisms that appeared in the Cambrian period, and with the extensive conservation of core genetic, cellular, and developmental processes that accompanied this diversity. From their analysis of the interdependence of such diversification and conservation they draw this principal conclusion: "As we look at breakthroughs in metazoan design since the pre-Cambrian, they seem to involve a succession of new attributes of evolvability, as if evolvability has itself evolved"—as if the major transitions of evolution depended upon the acquisition of ever more sophisticated motors of change.[51]

## What Is Life?

How do all these findings affect our thinking about genetics, development, and evolution? Needless to say, their full implications have yet to be explored, but already at least two lessons can be inferred with reasonable confidence. The first concerns the nature of genetic stability and the second the dynamics of genetic transformation and hence of evolutionary change. Taken separately, each of these implies a major reversal of one of the key expectations with which the century began. Taken together, they imply radical modifications in our view of the relationships that tie genetics, development, and evolution together. First, the question of stability.

By now, we have abandoned the hope of finding in the molecular structure of particulate genes an adequate explanation for the stability of biological organization across generations. We have learned that genetic stability is itself a consequence of biological organization, and while it may be a prerequisite for natural selection, the mechanisms for guaranteeing such stability are themselves an achievement of evolution. Furthermore, these mechanisms are not static but dynamic, and an explanation of how they do their job will have to be sought in the complex systems of cellular dynamics that are at one and the same time the products and the safeguard of genetic information.

What might such an explanation look like? One of the ironies of this history is that the very man who had been the source of the model leading Schroedinger astray gave us a clue. At a meeting in Paris in 1949 on mechanisms of genetic continuity—long after his quantum mechanical model of gene structure had failed—Max Delbrück sketched a quite different kind of model for achieving biological stability. As he explained, a system of cross-reacting and mutually inhibiting chemical reactions can lead to not just one stable steady state but to multiple steady states.[52] With such a mechanism, stability does not depend on the immutability of individual particles but solely on the dynamics of their interaction. Delbrück introduced this new model as an explicit alternative to arguments that had been presented for the existence of cytoplasmic genes—that is, as a nongenetic way to account for the stability of certain kinds of cellular inheritance, and not as a way to account for genetic stability as such.[53] And indeed, the primary use to which this model has subsequently been put has been to stimulate work on the stable steady states of metabolic regulatory networks. Yet it can also be seen (as it was by some) as harking back to a pre-Weismannian tradition in which heredity in general was regarded as a manifestation of biological regulation—in the words of David Nanney, as a "type of homeostasis."[54]

But homeostasis (or self-maintenance) has turned out to

be only one part of the story. The capacity to generate rapid change in times of stress suggests a far more dynamic mode of organization than anything suggested by Delbrück's model of steady states. Yet even so, that simple model may point us in the right direction, and perhaps especially so for thinking about Schroedinger's question, "What is life?" Historically, biologists have in fact been quite divided in their approach to an answer. Throughout Weismann's own century, the dominant tradition held that the defining feature of life was its organization. But after Weismann, a different tradition came to dominate, one which tended to define life in terms of genes and their replication. Still, a minority tradition has continued to look to self-maintaining (or autocatalytic) metabolic systems for the essence of life.

This divide is perhaps most conspicuous in debates about the origin of life. Which came first—genes or cells? Replication or self-maintenance? In 1985 Freeman Dyson, another physicist, published a small book called *Origins of Life*. Revisiting Schroedinger's question, he suggests that Schroedinger's approach exemplified a long-standing over-preoccupation with genes. Life, he argues, requires not just nucleic acid but also a metabolic system for self-maintenance; hence, the overwhelming likelihood is that it had not one but two origins. The emergence of living systems as we know them could have come about as a result of a symbiotic fusion between two independently evolved prior subsystems—one a rapidly changing set of self-reproducing but error-prone nucleic acid molecules and the other a more conservative autocatalytic metabolic system specializing in self-maintenance.

Over the last fifteen years, Dyson's picture of dual origins of life has gained increasing currency. According to this hypothesis, out of the interactions permitted by the conjoining of these two subsystems emerged a mechanism of heredity able to ensure the presence of a particular genotype over a period of time long enough for natural selection to begin its work and yet flexible enough to generate the variability on which natural selection could act. For then and only then would it be possible to evolve such exquisitely creative mechanisms as are needed for the genetic code, individuality, multicellularity, or sex. The rest, as they say, is history. But what a dynamic history it would have had to be.

## 2

## THE MEANING OF GENE FUNCTION: WHAT DOES A GENE DO?

Those were the genes, the living germs, bioblasts, biophores—lying there in the frosty night, Hans Castorp rejoiced to make acquaintance with them by name. Yet how, he asked himself excitedly, even after more light on the subject was forthcoming, how could their elementary nature be established? If they were living, they must be organic, since life depended upon organization. But if they were organized, then they could not be elementary, since an organism is not single but multiple. They were units within the organic unit of the cell they built up. But if they were, then, however impossibly small they were, they must themselves be built up, organically built up, as a law of their existence; for the conception of a living unit meant by definition that it was built up out of smaller units which were subordinate; that is, organized with reference to a higher form. As long as one spoke of living units, one could not correctly speak of elementary units, for . . . there was no such thing as elementary life, in the sense of something that was already life, and yet elementary.

THOMAS MANN, *The Magic Mountain* (1924)

On a number of occasions in the preceding chapter, I referred to genes that "lead to" or "are responsible for" a particular effect, and indeed it is difficult to find another form of expression. But what does a gene *do*? How does it "lead to" an effect, and in what sense is it responsible for that effect? In contemporary terms, the question takes the form: What is the relation between a gene's structure and its function? Indeed, much of today's research is devoted to this question, and it is yielding dramatic and unsettling results. For the results raise as many questions as they answer—so much so that we find ourselves obliged to return to that most elementary of questions: What, in fact, *is* a gene?

## A BRIEF OVERVIEW

Throughout the history of both classical and early molecular genetics,[1] the gene was generally assumed to be not only a fixed and unitary locus of structure and function but also a locus of causal agency. T. H. Morgan, for example, regarded the idea that genes are the causal agents of development as so basic and so self-evident that an understanding of heredity did not require its elaboration. He wrote: "The theory of the gene is justified without attempting to explain the nature of the causal processes that connect the gene and the characters."[2] Indeed, I have argued that just such an attribution of causal agency was implicit in the very notion of "gene action."[3] This phrase was the primary locution by which geneticists between the mid 1920s and the 1960s referred to "the causal processes that connect the gene and the characters," and it was a way of talking that at least tacitly granted to genes the power to act, even in the absence of any information about *how* they might act. This same way of talking endowed the gene with a most curious constellation of properties. At one and the same time, the gene was bestowed with the properties of materiality, agency, life, and mind.

The assumption that materiality (and especially particulate materiality) lent the gene its fixity, that is, the permanence it required as a stable unit of transmission, has already been discussed in Chapter 1. But endowing the gene with the power to act added to the property of materiality the further implication of agency. The capacity to reproduce itself—traditionally taken as the defining property of life—lent the gene vitality. And finally, attributing to the gene the capacity to direct or control development effectively credited it with a kind of mentality—the ability to plan and delegate. The net result was a gene with something of a Janus-faced quality. Part physicist's atom and part Platonic soul, it was assumed capable simultaneously of animating the organism and of directing (as well as enacting) its construction. As Erwin Schroedinger was later to put it, it was "law-code and executive power—architect's plan and builder's craft—in one."[4] Even to this day, the gene is sometimes referred to as "the cell's brain."[5]

In fact, however, the implicit attributions of vitality and mentality long preceded the notion of gene action, and probably even the term *gene* itself. It might therefore be more accurate to say that the discourse of gene action perpetuated an expectation that had already been built into the earliest conceptions of hereditary units. Indeed, it was just such an implication that had led Hugo de Vries to argue that these units were "not the chemical molecules; they are much larger than these and are more correctly to be compared with the smallest known organisms,"[6] or later, "even after more light on the subject was forthcoming" that had led Thomas Mann to conclude "there was no such thing as elementary life, in the sense of something that was already life, and yet elementary."

Yet the science of genetics depended on just such elementary units, and as that science grew, so too did the conviction of its practitioners that these units—being chemical molecules of one sort or another—do indeed have actual physical existence. Here they sharply distinguished themselves from their forebears. Whatever the gene might be, it would not serve to think of it as a minute organism. "It is inadmissable," wrote the Dutch geneticist Avend Hagedoorn in 1911, "to try to explain the facts of evolution and inheritance by the behaviour of living particles which have been invented simply to admit of this explanation."[7] Four decades later, looking back on the prehistory of genetics with the advantage of hindsight, the American geneticist H. J. Muller observed with some wonder: "Those who postulated 'pangenes', 'determinants', or other self-reproducing particles in the old days did not seem to realize what a monster they had by the tail. They were still, subconsciously, so close to the ancient lore of animism, in which practically all things were living . . . that to attribute reproduction to a particle . . . itself living anyway, seemed to present no problem."[8]

Certainly, no classical geneticist had been more acutely aware than Muller was of just how sizeable a problem it was to attribute self-reproduction to a particle. Nevertheless, for Muller as for others, the assumption that a gene was endowed with the inherent capacity to reproduce itself stood firm. (Muller referred to it as its "specific autocatalytic power.") Moreover, genes needed not only autocatalytic power but also "heterocatalytic" power: they needed also the ability "to act as the determiners of the properties of the cell and of the organism as a whole."[9] The question was how: How do they reproduce themselves? And how do they "act as the determiners of the properties of the cell"? What is it that a gene *does*? How, in short, do genes *function*?

By any account, this had to be a central question for genetics, but of necessity it was a question for the future. For now, Muller and his colleagues had only their expectations. Yet they were sustained by the confidence that the solution to this problem of gene function would be found in the molecular structure of the gene's defining chemical constitution. The task that lay ahead for geneticists was thus obvious, and it was twofold: they needed not only to find the structure of a gene but also to explain how such a structure, whatever it might be, could translate into function. Above all, they needed to explain how it is that the unit of hereditary, as de Vries had earlier put it, "impresses its character upon the cell."[10]

But what a formidable task! The problem of gene structure was challenging enough, and historians have written extensively on the difficulties it posed.[11] But the problem of gene function was more challenging by far; indeed, to many biologists, an answer to the question of how structure could translate into function seemed so difficult to imagine as to virtually defy reason. What kind of chemical molecule could fill so demanding a role, serving not only to preserve genetic memory through the generations but also, in each generation, to steer the course of individual development? Some suggested that enzymes might do the trick; Leonard Thompson Troland was one of the first to make such an argument. In fact, he went so far as to claim that "On the supposition that the actual Mendelian factors are enzymes, nearly all [the biological enigmas] instantly vanish."[12] But others (not surprisingly) remained skeptical. Was such an answer really so different from the hypothesis of genes as organisms? Some thought not.

To J. S. Haldane—to take just one example—the supposition that genes were "so constituted physically and chemically that . . . [they] gave rise to all the amazingly specific details of the structure and activity observed in the adult organism" amounted to little more than a kind of preformationism—"in reality only a variant of the 'box-within-box' theory, an extremely complicated molecular structure capable of producing the adult form being substituted for the original miniature adult."[13] But J. S. Haldane was not a geneticist (unlike his son, J. B. S. Haldane), and while similar complaints were voiced by many others, they too tended to arise from other biological disciplines—from physiology or embryology. And for the most part there they remained.

In the 1930s, genetics was still a fledgling discipline seeking to establish its position within the hierarchy of biological sciences, and its practitioners had little interest in hearing about what it could not do. As a consequence, criticisms like Haldane's did little to slow the progress of genetics itself, and, one might say, fortunately so. Indeed, one of the great virtues of the discourse of gene action was that it permitted geneticists to pursue their research programs so productively, and for so long, without even a glimmer, anywhere on the horizon, of an answer to the question of *how* genes act. That question might be fundamental, but as long as no one could envision a path that might lead to its resolution, dwelling on it was probably a waste of time. Other questions—problems that were possible to address with the tools at hand—were abundant, and the simple expression "gene action" served as an effective way to bracket what could not be explained.

The break-through came in the middle of the century. The moment that is sometimes celebrated as marking the dawn of the new era came in the early 1940s, with George Beadle and Edward Tatum's formulation of the one gene–one enzyme hypothesis. It grew primarily out of their work on mutants of *Neurospora* (a fungus), in which the authors succeeded in tying specific mutations to the failure of specific steps in a metabolic pathway. This, they argued, demonstrated that genes control biochemical reactions. But neither Beadle nor Tatum had any notion of the physical or chemical means by which such genetic control could be effected.

What really put teeth into the one gene–one enzyme hypothesis, and hence into the notion of gene action, was the identification of the genetic material with DNA and, in 1953, the deciphering of DNA's double helical structure. After the publication of Watson and Crick's first paper in April 1953, things moved with lightning speed. A mere five weeks later, Watson and Crick published a second paper—on "Genetical Implications of the Structure of Deoxyribonucleic Acid"—whose main point was to argue that the structure of DNA shows us "how it might carry out the essential operation required of a genetic material, that of exact self-duplication."[14] But at the same time, they also noted the large number of permutations that would be possible in a long molecule and, with that observation, the likelihood "that the precise sequence of the bases is the code that carries the genetical information." Within such a framework, the one gene–one enzyme hypothesis took on a new kind of sense. Now it could be understood as suggesting a direct correspondence between the sequence of nucleotides in a gene and the sequence of amino acids in a protein (see Figure 4).

Experimental demonstration of this new reading of the one gene–one enzyme hypothesis (that is, of the colinearity of genes and proteins) was the great challenge of the next few years, and its pursuit drove many if not most of the remarkable achievements of the new molecular biology. Seymour Benzer approached the problem by constructing a fine-structure map of mutations in a particular gene *(rII)* of the bacteriophage T2, in which a single mutation might reflect an alteration of a single nucleotide. His work provided the crucial genetic evidence for the linearity of the internal structure of genes. From the other side, biochemists were hard at work developing methods to identify the particular amino acid at fault in a mutant protein.

By 1957 sufficient evidence was at hand for Francis Crick to go public with an explicit formulation of the "sequence hypothesis": "In its simplest form," he wrote, "it assumes that the specificity of a piece of nucleic acid is expressed solely by the sequence of its bases, and that this sequence is a (simple) code for the amino acid sequence of a particular protein."[15] As for finding the actual code by which the one sequence is translated into the other, what proved most effective in the end was a direct biochemical approach.[16] Marshall Nierenberg and Heinrich Matthaei made the first definitive association between the two different kinds of sequences in 1961: they showed that a uniform stretch of nucleic acid consisting of a single nucleotide (uridine) leads to the test tube synthesis of a polymer string made up of only one amino acid (phenylalanine). By 1966, using similar biochemical protocols, molecular biologists succeeded in estab-

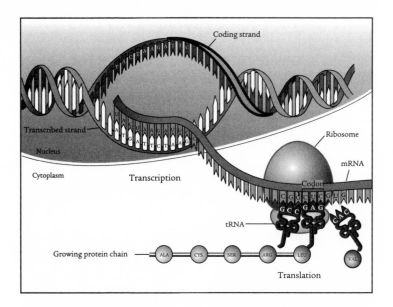

**Figure 4: The Central Dogma: transcription and translation.** To make a protein molecule, the DNA double helix separates at the site of a gene, and transcribing enzymes (not shown) copy the bottom strand of nucleotides into a complementary mRNA strand. Where there is a G in the DNA, a C appears in mRNA; where there is a C in DNA, a G appears in mRNA; where there is a T, an A appears. However, an A in DNA appears as a U instead of a T in mRNA. Consequently, the upper or coding strand of the DNA has the same sequence as the mRNA except that T is present in DNA and U in mRNA. After the mRNA is transported from the nucleus, it joins ribosomes in the cytoplasm, where it is translated. Each codon (or triplet of bases) in the mRNA is complementary to a specific transfer RNA (tRNA), and each tRNA carries a specific amino acid to add to the growing protein chain. In this example, the amino acids arginine, leucine, and valine are being added to the chain, in the order dictated by the codons in the mRNA. When the chain is completed, it will fall off the mRNA-ribosome complex and become a functioning protein molecule. *(By Kathy Stern, from D. G. Nathan, Genes, Blood, and Courage: A Boy Called Immortal Sword, 1995; reproduced by permission of Harvard University Press.)*

lishing a complete "genetic code" by which successive trip-
lets of nucleotides could be translated into a sequence of
amino acids.

The completion of the code marks the culmination of
what Hans-Jorg Rheinberger has described as the "preco-
cious simplicity" of the new molecular genetics.[17] At long
last we had an answer to the question of how genes act.
What does a gene do? It makes (or encodes) an enzyme.
Moreover, a defect in the gene leads to a defect in that enzyme (or perhaps its absence), which in turn can often be directly correlated with the abnormality (or absence) of a particular trait. The answer was stunning in its simplicity; also,
it had the elegance of a mathematical equation. Perhaps
more importantly yet, it fulfilled the long-standing expectation that the function of a gene could be read in its structure, if only we knew how to decipher that structure. Now
all that was needed was a way to get at the sequence of nucleotides in the DNA.

Such a wonderfully simple picture could scarcely fail to
capture the imagination, and capture the imagination it
did. It established DNA as the molecule that not only holds
the secrets of life but that also executes its cryptic instructions—it was, in short, the "Master Molecule." In the col-
loquial paraphrase of the "central dogma" formulated by
Francis Crick in 1957, "DNA makes RNA, RNA makes protein, and proteins make us." This picture has reigned su-
preme ever since, among scientists, students, and lay readers
alike, and it has spawned a plethora of associated images.
We think of the cell's DNA as the genetic program, the lin-
gua prima, or, perhaps best of all, the book of life. Indeed,
the metaphor of DNA as the book of life has become a

media favorite ever since the start of the Human Genome Project.

Yet even at its highest moment, this immensely satisfying picture was already beginning to show a few blemishes. For a long time, these appeared as nothing more than minor wrinkles, hardly worth noticing against the astonishing beauty of the whole. But over the last two decades, and especially since the rise of genomics, some of these minor wrinkles have grown into major chasms. As every molecular biologist now knows, the secrets of life have proven to be vastly more complex, and more confusing, than they had seemed in the 1960s and '70s. In the remainder of this chapter, I will review some of the complications that had already begun to accrue in the early days of molecular biology—before genomics, before the move to higher organisms, even before the code had begun to be deciphered—and attempt to chart their staggering proliferation over the years since.

REGULATOR GENES

The first wrinkle on the face of the central dogma came in 1959, when François Jacob and Jacques Monod introduced a distinction between "structural genes" and "regulator genes."[18] From their studies of bacterial adaptation, they concluded that an understanding of the biosynthesis of proteins requires the assumption that chromosomes house more than one kind of gene—not only genes that code for the proteins needed to make an organism (structural genes) but also genes that regulate the rate at which these structural genes are transcribed. By 1961 their combined genetic and biochemical analyses led to the identification of just such a new entity—"a 'regulator gene' as a hereditary determinant which, in its active state, controls the rate of transcription of certain specific structural genes without itself contributing any structural information to the proteins."[19] Summarizing the implications of their findings, they wrote, "The purely structural (one gene–one enzyme) theory does not consider the problem of gene expression. The discovery of a new class of genetic elements, the regulator genes, which control the *rate* of synthesis of proteins, the *structure* of which is governed by *other* genes, does not contradict the classical concept, but it does greatly widen the scope and interpretative value of genetic theory."[20]

In fact, one might say that *gene expression* had all along been the Achilles' heel not just of the one gene–one enzyme hypothesis but of the very notion of gene action. Even T. H. Morgan, one of the most forceful proponents of that notion, had recognized the problem as soon as he turned his attention back to the study of embryology. Writing almost three decades before Jacob and Monod's analysis of regulation, Morgan had this to say: "The implication in most genetic interpretation is that all the genes are acting all the time in the same way. This would leave unexplained why some cells of the embryo develop in one way, some in another, if the genes are the only agents in the results. An alternative view would be to assume that different batteries of genes come into action as development proceeds."[21] But how do these "different batteries of genes come into action"? This view of development implies the participation of other agents, not just the genes—agents whose function is to call particular genes into action at the appropriate time and place.

Jacob and Monod were not studying the complex development of embryos but the regulation of protein synthesis in lowly bacteria.[22] Yet even in bacteria, they had found a way to resolve Morgan's dilemma. Genes do not simply *act*: they must be *activated* (or inactivated), and it is here, in the activation (or inactivation) of structural genes, that regulatory genes come into play. Still, a key difference can be discerned between Morgan's solution and theirs, and that difference lay in the attribution of agency. Morgan had suggested that the genes might not be the only agents; but in Jacob and Monod's formulation, whatever structural genes might lack in effective agency would be provided by another kind of gene, the "regulator" gene.

The concept of regulator genes lay at the core of Jacob and Monod's immensely influential model for the regulation of gene activity (the operon model), but regulator genes constituted only one part of the apparatus they proposed. The term *operon* refers to a linked cluster of regulatory elements and structural genes whose expression is coordinated by the product of a regulator gene situated elsewhere in the genome. The role of the regulator gene was to "provoke the synthesis" of a repressor, which in turn regulates the transcription of the structural genes by binding to an operator region adjacent to the structural genes.[23] The term *operator* refers to yet another genetic element, one that is equally critical to regulation even though it has not yet been called a gene (see Figure 5).

In the years since Jacob and Monod proposed their operon model, and especially since molecular biologists began to study gene regulation in higher organisms, regulatory elements in the genome that do not encode structural proteins, and may not encode regulatory proteins—indeed,

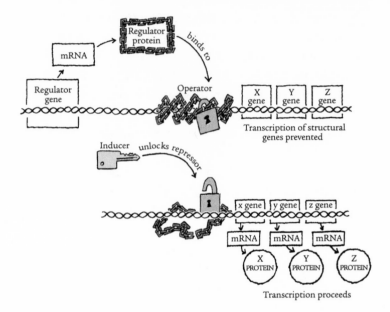

**Figure 5: The operon model for gene regulation.** Gene transcription can be switched on and off by regulator proteins. In the operon model, a regulator protein (repressor) binds to the operator, preventing RNA polymerase from transcribing the structural genes. Gene expression is now turned OFF. When the inducer is present, it releases the repressor from the operator by directly binding to, and thereby changing the shape of, the protein. As long as the operator remains free of the repressor, RNA polymerase can proceed in the transcription of the operon's structural genes into mRNA. Gene expression is now turned ON. *(By Nick Thorkelson.)*

may not even be transcribed—have proliferated extravagantly both in number and kind. Some are *promotor* and *terminator* sequences; others are *leader* sequences. Still others are *activator* elements, located either upstream or downstream from the gene to be activated. And so on. Many of these may be considered genes in the sense that they provide templates for a gene product, but others do not "make" anything: their function is merely to provide specific sites at which other proteins of the right kind and in the right configuration may attach themselves (bind) to the DNA. Indeed, current estimates suggest that only about three percent of the human genome codes for amino acid sequences, and in other organisms the percentage may be even less. Thus, when we read that the yeast genome has 6,200 genes, are we to understand this number as including regulator genes as well as structural genes?[24] And what about those genetic elements that provide only binding sites? Or should these be considered parts of the structural or regulatory genes they regulate? If so, where would we locate such a gene? Often, these elements are scattered far from the coding sequences they regulate. What then should we count as the beginning and end of a gene?

### Splicing and Editing the Message

An even bigger monkey wrench was thrown into the concept of the gene in the late 1970s, when Richard Roberts and Phillip Sharp discovered split genes. Many of the genes that code for proteins in higher organisms turn out not to be continuous but fragmented—composed of coding (or expressed) segments of DNA *(exons)* interspersed with long noncoding regions *(introns)* that were presumed, at least at first, to have no function at all.[25] Intronic DNA was commonly referred to as "junk DNA," and there seemed to be a great deal of it.[26] The presence of junk DNA posed an obvious problem not just for biologists (What is it doing there, and why has natural selection allowed it to persist?) but, it seemed, for organisms as well (How do the mechanisms for protein synthesis proceed with so much junk DNA around?).

For genes that are fragmented in this way, there is no strict one-to-one correspondence between the sequence of a gene and that of the protein it gives rise to. Thus the original RNA transcript directly transcribed from the gene (the messenger RNA, or mRNA) must be processed to remove these junk sequences before protein synthesis can begin. As it is now understood, a number of special protein/RNA complexes (called *spliceosomes*) prepare the primary transcript for protein synthesis by excising the introns. The remaining exons are then spliced together to form a continuously coding (mature) transcript.

But exons can be spliced together in more than one way, and this flexibility is often put to good use in developing organisms. *Alternative splicing* is the term biologists use to refer to the construction of different mRNA transcripts from a single primary transcript; these different mature transcripts, in turn, lead to the synthesis of correspondingly different proteins. As many as one third of eukaryotic genes are routinely subjected to such variable readings, where the decision as to how the primary transcript is to be read is itself carefully regulated, depending on the state and type of the cell.[27] Just how the decision to splice together one transcript rather than another is regulated is still under investigation, but a variety of proteins characteristic of particular cell types have been shown to influence the pattern of splicing by binding to specific sites on the primary transcript.

The bottom line is that, depending on the context and stage of development of the organism in which a primary transcript finds itself, different pieces of the transcript may be cut and pasted together to form a variety of new templates for the construction of a corresponding variety of proteins. Additional variation may be generated, first, by the presence of more than one site on the primary transcript at which the mature transcript can begin and, second, by the presence of multiple sites at which the primary transcript can be spliced. Finally, to complicate matters further, the very distinction between introns and exons seems not to be fixed. In some cases, the synthesis of still other proteins has now been shown to derive from stretches of intronic DNA—that is, from regions that had originally been consigned to the pile of "junk" DNA, but, as we now know, mistakenly so (see Figure 6).[28]

How many different proteins can be synthesized from the same primary transcript? The number varies greatly from one organism to another, and estimates seem to increase daily. Two decades ago, splicing variants could be identified only through laboratory analyses of mRNA transcripts and proteins, but nowadays potential variants can be read directly from the sequence data provided by the Human Genome Project. Accordingly, the number of different proteins at least hypothetically associated with a particular gene has escalated sharply, and in some organisms that number now reaches into the hundreds.[29]

Moreover, alternative splicing is not the only way in which variation can be generated from messenger RNA. RNA transcripts are subject to a variety of other kinds of editing as well, equally systematic and equally well-regulated. For example, in some organisms, mature transcripts can be formed by splicing together exons from two different primary transcripts. Or still more spectacularly, even the spliced transcript may later be modified by insertion of for-

(a) Alternative selection by promoter site

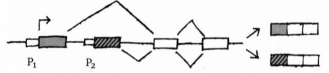

(b) Alternative selection of cleavage/polyadenylation sites

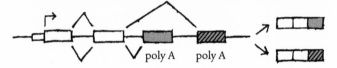

(c) Intron retaining mode

(d) Exon cassette mode

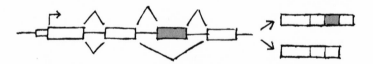

**Figure 6: Variations on alternative splicing.** Different ways of generating variability from splicing. *(After G. M. Malacinski and D. Freifelder, Essentials of Molecular Biology, 1998, Jones and Bartlett Publishers.)*

eign bases or by the replacement of one base by another, thereby giving rise to proteins for which no corresponding coding sequence exists in the DNA. The question that can no longer be deferred is obvious: Which of these different transcripts corresponds to what we should call the gene?

So far, I have been using the term *gene* to refer to the original stretch of DNA. But doing so means that we have to give up on the notion, even for structural genes, that one gene makes one enzyme (or protein). One gene can be employed to make many different proteins, and indeed the expression "one gene–many proteins" has become fairly common in the literature. The problem with this formulation is that the gene has lost a good deal of both its specificity and its agency. Which protein should a gene make, and under what circumstances? And how does it choose?

In fact, it doesn't. Responsibility for this decision lies elsewhere, in the complex regulatory dynamics of the cell as a whole. It is from these regulatory dynamics, and not from the gene itself, that the signal (or signals) determining the specific pattern in which the final transcript is to be formed actually comes. Unraveling the structure of such signaling pathways has become a major focus of contemporary molecular biology, and while the temptation remains strong to order these pathways as linear sequences of events deriving from the action of yet other genes, the evidence that is accumulating makes such a simple ordering ever more difficult.[30]

An obvious alternative would be to consider the mature mRNA transcript formed after editing and splicing to be the "true" gene. But if we take this option (as molecular biologists often do), a different problem arises, for such genes exist in the newly formed zygote only as possibilities, designated only after the fact. A musical analogy might be helpful here: the problem is not only that the music inscribed in the score does not exist until it is played, but that the players rewrite the score (the mRNA transcript) in their very execution of it. Furthermore, such genes have none of the permanence traditionally expected of genes—these recompiled mRNA transcripts are called into being only as needed and generally have rather short lifetimes. Indeed, they do not reside on the chromosome and, in some cases, might not even be found in the nucleus—that is, the final version of the transcript may be put together only after the original transcript has entered the cytoplasm (see Figure 7).[31]

## REGULATION OF PROTEIN FUNCTION

The discovery of complex modes of RNA editing has greatly confounded the prospects of a simple relation between genes and proteins, but the difficulties in understanding gene function do not end with the regulation of protein synthesis. In some ways, the synthesis of a protein marks only the beginning of the story of gene function. The rest of the tale centers on the function of the protein and the ways in which that function is regulated.

The assumption that the function of a gene has been identified once the amino-acid sequence of the protein is determined ignores the well-established fact that a protein can function in many different ways, depending on its context. The first evidence that protein function is itself subject to cellular regulation came in 1963 with the discovery of allosteric effects, and from the same group that had given us the operon model two years before.[32] Proteins too have their regulatory sites, and the term *allostery* refers to changes in the three-dimensional structure of a protein that can be induced by the binding of certain other *(effector)* molecules at these sites; in turn, change in conformation alters the function and activity of the protein. Allostery thus names a particular mechanism for regulating protein function, and over

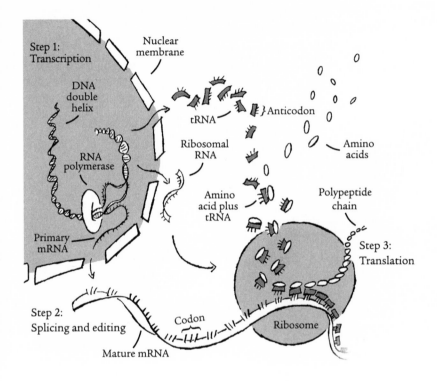

**Figure 7: Protein synthesis.** Protein synthesis can be divided into three parts: (1) Forming the primary transcript (transcription): One strand of the DNA double helix is used as a template by the RNA polymerase to synthesize a messenger RNA (primary mRNA). (2) Splicing and editing: Before the synthesis of a protein begins, the corresponding primary mRNA migrates from the nucleus to the cytoplasm. During this process, the primary mRNA is transformed by splicing and editing into a mature mRNA ready for translation into a sequence of amino acids. (3) Translation: The ribosome binds to the mRNA at the start codon (AUG) that is recognized only by the initiator tRNA. The ribosome proceeds to the elongation phase of protein synthesis. During this stage, complexes, composed of an amino acid linked to tRNA, sequentially bind to the appropriate codon in mRNA by forming complementary base pairs with the tRNA anticodon. The ribosome moves from codon to codon along the mRNA. At the end, a release factor binds to the stop codon, terminating translation and releasing the complete polypeptide from the ribosome. *(By Nick Thorkelson.)*

the thirty years since its first description it has been found to operate in a wide range of proteins.

Furthermore, research over the same period has led to the identification of numerous other mechanisms that regulate protein function. As a consequence, molecular biologists have become obliged to adjoin to their updated notion of one gene–many proteins an additional admonition: one protein–many functions. Constance Jeffery refers to proteins with a multiplicity of functions as "moonlighting proteins," and she writes, "The multiple functions of such moonlighting proteins add another dimension to cellular complexity and benefit cells in several ways. However, cells have had to develop sophisticated mechanisms for switching between the distinct functions of these proteins."[33]

How do we know that such higher order mechanisms have in fact evolved? The answer is simple: their existence is attested to by the manifest orderliness and precision with which organisms typically proceed through their various developmental stages, despite the many options available and despite the many opportunities for making "wrong" choices.

## A CONCEPT IN TROUBLE

Fifteen years ago, the historian and philosopher of biology Richard Burian observed, "There is a fact of the matter about the structure of DNA, but there is no single fact of the matter about what the gene is."[34] In the interim, things have only gotten worse. As a result of the phenomenal progress that has been made in the identification, mapping, and sequencing of particular genes, we have learned a tremendous amount about the structure and function of genetic material, and much of what we have learned falls outside the frame of our original picture. The complications brought by the new data are vast; those I have discussed in this chapter are merely the tip of the iceberg. Taken together, they threaten to throw the very concept of "the gene"—either as a unit of structure or as a unit of function—into blatant disarray.

Techniques and data from sequence analysis have led to the identification not only of split genes but also of repeated genes, overlapping genes, cryptic DNA, antisense transcription, nested genes, and multiple promotors (allowing transcription to be initiated at alternative sites and according to variable criteria). All of these variations immeasurably confound the task of defining the gene as a structural unit.

Similarly, discovery of the extensive editorial process to which the primary transcript is subject, of regulatory mechanisms operating on the level of protein synthesis, and others operating even on the level of protein function confound our efforts to give a clear-cut functional definition of the gene. As Peter Portin observes, "Our knowledge of the structure and function of the genetic material has outgrown the terminology traditionally used to describe it. It is arguable that the old term gene, essential at an earlier stage of the analysis, is no longer useful."[35] William Gelbart, working at the forefront of molecular genetics, concurs in suggesting that the gene might be "a concept past its time." "Unlike chromosomes," Gelbart writes, "genes are not physical objects but are merely concepts that have acquired a great deal of historic baggage over the past decades." To be sure, the concept of the gene has played a crucial role in leading us to our present understanding of biological phenomena, but today, he suggests, "we may well have come to the point where the use of the term 'gene' . . . might in fact be a hindrance to our understanding."[36]

There is more than a little irony in this present state of affairs, for never in the history of the gene has the term had more prominence, in both the scientific and the popular press. Daily, we are informed of the identification of new "disease-causing" genes, with the list of corresponding "genetic" diseases growing steadily longer.[37] Similarly, we are told that much of human behavior that had previously been thought to be voluntary, or culturally induced, is a product of our genes. Certainly, astonishing progress has been made in understanding the importance of genetic mutations in the incidence of many diseases (including a number of behavioral disorders). A number of conditions have now been definitively tied to mutations in specific genes. The simplest and most clear-cut cases are the single-gene disorders (Tay-Sachs, Huntington's disease, cystic fibrosis, thalassemia, and phenylketonuria [PKU], among others). Such examples remain rare, however, and even in these clear-cut cases much remains to be understood about the processes that link the defective gene to the onset of disease.

In conditions that are known to involve the participation of many genes (such as certain kinds of heart disease, stroke, psychoses, diabetes), the limits of current understanding are far more conspicuous. The net effect is that, while we have become extraordinarily proficient at identifying genetic risks, the prospect of significant medical benefits—benefits that only a decade ago were expected to follow rapidly upon the heels of the new diagnostic techniques—recedes ever further into the future. As D. J. Weatherall, Director of the Institute of Molecular Medicine at the University of Oxford, writes, "Transferring genes into a new environment and enticing them to . . . do their jobs, with all the sophisticated regulatory mechanisms that are involved, has, so far, proved too difficult a task for molecular geneticists."[38] Part of the difficulty, of course, is in understanding what it is that genes *do*.

In other words, behind the so-called therapeutic gap between genetic screening and medical benefits lies the complexity of the regulatory dynamics that now puts the very concept of the gene into jeopardy. Indeed, "irony" may be far too weak a word to capture the incongruities of our current situation. For the basic fact is that, at the very moment in which gene-talk has come to so powerfully dominate our biological discourse, the prowess of new analytic techniques in molecular biology and the sheer weight of the findings they have enabled have brought the concept of the gene to the verge of collapse. What is a gene today? As we listen to the ways in which the term is now used by working biologists, we find that the gene has become many things—no longer a single entity but a word with great plasticity, defined only by the specific experimental context in which it is used.

It can of course be argued, as Rheinberger and other historians remind us, that the concept of the gene never was "either unitary, comprehensive, or clean."[39] Nor, for that matter, has it been stable over the course of its history. Yet there is, I suggest, one feature that clearly distinguishes the present from the past. Throughout the many variations and transformations that we have seen in the concept of the gene over the course of its lifetime, it had always been possible in the past to contain whatever definitional difficulties had plagued that concept; one might even say that it had been functional, both experimentally and professionally, to keep its internal incoherence under wraps. What is distinctive today is that progress in molecular biology has now made it possible to break this historic silence.

And for this development, the newly available sequence information has been especially valuable. As the original goals of the Human Genome Project approach fulfillment, those who hoped that knowledge of the sequence of the genome would suffice for understanding the organism may be disappointed. But in helping to reveal the naiveté of those hopes and thereby to set us on a more realistic track toward understanding how organisms develop, function, and evolve, the contributions of HGP have been invaluable. Clear and demonstrable gaps have now been exposed between the many different attributes that had historically been assumed to inhere in one single entity, the gene. To be sure, many different kinds of research have played their parts in the exposure of such gaps, but the role of the new sequence data has been of unmistakable importance.

Yet the lesson comes hard. Ever since the term *gene* was first introduced, confidence in the physical reality of the gene has always been accompanied by the assumption that structure, material composition, and function were all properties of one and the same object, be it a bead on a string or a stretch of DNA. Today, it is precisely that self-identity which has been disrupted. We have learned not only that function does not map neatly onto structure but also that function must be distinguished from a particular and pre-specified locus of the chromosome. To the extent that we can still think of the gene as a unit of function, that gene (we might call it the *functional gene*) can no longer be taken to be identical with the unit of transmission, that is, with the entity responsible for (or at least associated with) intergenerational memory. Indeed, the functional gene may have no fixity at all: its existence is often both transitory and contingent, depending critically on the functional dynamics of the entire organism.[40]

The moral of this chapter thus closely parallels that of Chapter 1. There, we learned that the source of genetic stability was not to be found in the structure of a static entity but that stability is itself the product of a dynamic process. Here, we learn that gene function needs also to be understood in dynamical terms. Because biological function inheres in the activity of proteins rather than of genes, the breakdown of the one gene–one protein hypothesis critically undercuts the possibility of attributing function to the structural unit that has traditionally been taken as the gene. Yet reconceived as a functional unit (for example, the spliced and edited mRNA sequence), the gene can no longer be set above and apart from the processes that specify cellular and intercellular organization. That gene is itself part and parcel of processes defined and brought into existence by the action of a complex self-regulating dynamical system in which, and for which, the inherited DNA provides the crucial and absolutely indispensable raw material, but no more than that.[41]

In short, the evidence accruing over recent decades obliges us to think of the gene as (at least) two very different kinds of entities: one, a structural entity—maintained by the molecular machinery of the cell so that it can be faithfully transmitted from generation to generation; and the other, a functional entity that emerges only out of the dynamic interaction between and among a great many players, only one of which is the structural gene from which the original protein sequences are derived. Or, to put it just a little differently, the function of a structural gene depends not only on its sequence but, as well, on its genetic context, on the chromosomal structure in which it is embedded (and which is itself subject to developmental regulation), and on its developmentally specific cytoplasmic and nuclear context.[42]

Just how the many other players—including regulatory sequences found elsewhere on the genome, the products of many other structural and regulatory genes, the complex signaling network of the living cell—are organized into a well-functioning and reliable whole is the question that dominates the attention of molecular geneticists today. We might even put today's question as a paraphrase of the one Howard Pattee, writing in a very different context, posed in 1969: How does a sequence become a gene?[43]

The reader will surely have noticed how cumbersome all of this has become. One reason is that the story itself has become so complicated; but another reason may be that the very use of the term *gene* has become an impediment to its exposition. In 1911, when Johannsen first introduced his "applicable little word" to an English-speaking audience, he believed that new terminology was made necessary by the fact that older terms had become dysfunctional, exerting too powerful a hold on our thinking: "It is a well-established fact," he wrote, "that language is not only our servant, when we wish to express—or even to conceal—our thoughts, but that it may also be our master, overpowering us by means of the notions attached to the current words."[44] Perhaps we are in a comparable situation today. Perhaps it is time we invented some new words.

## 3

## The Concept of a Genetic Program:
## How to Make an Organism

The time has come for direct attack upon the central problem of biology, the problem of how it is that a single cell, the fertilized egg, gives rise to an adult creature made of many different kinds of cells. This process, which we know as development, has been described and thought about by biologists for as long as there has been a science of biology. Its nature has remained a mystery because we have not heretofore understood enough about the nature of life itself. Today we do. We know that all cells contain the directions for all cell life, written in the DNA of their chromosomes, and that these directions include specification of how to make the many kinds of enzyme molecules by means of which the cell converts available substances into metabolites suitable for the making of more cells. This is the picture of life given to us by molecular biology and it is general, it applies to all cells of all creatures. It is a description of the manner in which all cells are similar. But higher creatures, such as people and pea plants, possess different kinds of cells. The time has come for us to find out what molecular biology can tell us about why different cells in the same body are different from one another, and how such differences arise.

JAMES BONNER, *The Molecular Biology of Development* (1965)

‖ ‖ ‖ ‖ ‖ ended Chapter 2 with a plea for some new words, but I neglected to mention one of the most important linguistic innovations of molecular biology, namely, the term "genetic program." Its introduction marked a crucial distinction between classical and molecular genetics, and it has enjoyed enormous popularity ever since. Indeed, the genetic program has come to be widely regarded as a fundamental explanatory concept for biological development—if not *the* fundamental concept. What exactly *is* a genetic program? And in what sense can it be said to explain development?

Here, as in Chapter 2, my concern is with the question of function, only now my focus is on the function of the genome as a whole, rather than on the function of individual genes. The question before us is not about the making of an enzyme but about the making of an organism. For even were we able to hold to a simple correspondence between one gene and one protein, we would still have to bridge the gap between proteins and organism: How can an organism be built out of the mere accumulation of different proteins? As we will see, once again the question hinges on our understanding of regulation, but this time around I want to focus more directly on the sense of agency that tends to inhere in the very notion of regulation (that is, in the supposition of a supervisory body responsible for regulating). Where, I ask, might such agency be located?

Yet even to talk about regulation is to jump ahead of the story. We need first to ask: How did geneticists think about the making of an organism prior to the introduction of the concepts of genetic programs and genetic regulatory mechanisms? To answer that question, we need to look at the terms and concepts with which earlier generations of geneticists were obliged to work, starting once again with the notion of gene action.

## From "Gene Action" to "Gene Activation"

The tacit implication of the discourse of gene action was not simply that action of an individual gene would lead to the formation of an individual trait (a "character") but that the development of an organism could be envisaged as a kind of summation of the action of many different genes. One should not expect to find such an assumption formulated explicitly; yet Alfred H. Sturtevant's classic rephrasing of the problem of embryogenesis comes close. Sturtevant, a geneticist from Morgan's group, presented a paper on "The Developmental Effects of Genes" at the 1932 International Congress of Genetics, and he began with this observation: "One of the central problems of biology is that of differentiation—how does an egg develop into a complex many-celled organism? That is, of course, the traditional major problem of embryology; but it also appears in genetics in the form of the question, 'How do genes produce their effects?'"[1] Moreover, he argued, genetics provides a way to answer this question: "It is clear that in most cases there is a chain of reaction between the direct activity of a gene and the endproduct that the geneticist deals with as a character." The task of the geneticist is therefore clear. It is to analyze these "chains of reaction into their individual links."[2] Doing so would enable us to answer the question of how genes produce their effects and, accordingly, to resolve the core problem of embryology. For presumably, if the direct activity of a single gene leads by some "chain of reaction" to the formation of a single trait, then a complete set of genes ought to lead to a whole organism.

At a bare minimum, such a presumption conspicuously ignores the effects of gene interaction, and for this reason Ernst Mayr famously dubbed this way of thinking "bean-bag genetics."[3] Mayr's critique was directed at population genetics, but the shortcomings of bean-bag genetics were even more evident to developmental biologists. Long before Mayr's critique, Morgan had cautioned his colleagues that an account based merely on the cumulative effects of gene action would not suffice for explaining development. Already in 1934 it was clear to him that some supplementary assumption was needed—for example, that genes act variably, being called into action at different times of development by other factors that might themselves be nongenetic. As it happened, however, Morgan's proposal of differential gene activation did not begin to take root among geneticists until the mid 1950s.

For some researchers, the impetus of greatest significance came from the cytological work of Wolfgang Beermann and his colleagues at the Max Planck Institute for Marine Biology in the early 1950s. Twenty years before, microscopic studies had revealed the presence of distinct bands on the giant salivary gland chromosomes of *Drosophila*, and at the time the appearance of such bands was taken as powerful confirmation of the physical reality of genes. Now, by observing variations in the structure of similar giant (or polytene) chromosomes across a number of different tissues of the genus *Chironomus*, Beermann and his colleagues were able to provide direct visual evidence for structural modifications (or "puffing") of individual loci that appeared to be specific to the tissue in question.

In 1956 Beermann presented his results to the American genetics community at the Cold Spring Harbor Symposium, explaining: "Puffing obviously indicates changes, most probably increases, in the activity of gene loci. Hence nuclear differentiation in cells of different function would be characterized by different specific patterns of activation along the chromosomes."[4] Here, finally, was unmistakable evidence that genes did not simply act but were subject to differential activation, and the relevance of this fact for the relation between genetics and development was obvious. And perhaps especially so to anyone already persuaded of the importance of developmental context for understanding gene action.

At that time, C. H. Waddington was one of the leading spokesmen for a developmental perspective on genetics. Since the 1930s his work had been directed toward building bridges between embryology and genetics, and when Beermann's findings were published, Waddington was among the first to take notice. In a 1954 essay on the physiology of development, he wrote: "The basic fact which we have to try to understand is that different cells in the body, although presumably containing the same genes, yet differentiate into quite different tissues. The fundamental mechanism must be one by which the different cytoplasms, or oöplasms, which characterize the various regions of the egg, act differentially on the nuclei so as to encourage the activity of certain genes in one region, of other genes in other places. Such specific activation of particular genes at certain times and places can actually be observed visually in favourable cases, for instance, in the important work of Pavan (1954), Mechelke (1953) and Beermann (1952) on the polytene chromosomes in various tissues of chironomids. The fact of differential activation of genes is, then, scarcely in doubt."

The question is: How might we envisage such a process? About this, Waddington noted, "there has been as yet little discussion." He did, however, make a proposal for where one might look: "There are innumerable different types of kinetic system which might be supposed to be in operation," and by way of giving a more definite picture he proposed an "exploration of the simpler varieties of these."[5] In short, he argued that looking at the steady states of interacting metabolic process (of much the same sort as Max Delbrück had earlier proposed as an alternative to the hypothesis of cytoplasmic genes) might give us some insight as to "the kind of system with which we are confronted."[6]

But Waddington's influence in the 1950s was rather limited, and especially among American geneticists.[7] Indeed, it might even be said that his contributions to developmental genetics are better appreciated today than they were then. In his lifetime, Waddington was something of an outlier. With one foot in genetics and the other in embryology, he had never belonged to the mainstream of either discipline; nor was he a participant in the new field of molecular biology. Moreover, as a committed follower of Alfred North Whitehead, he was a perennial critic of what he referred to as "the genetical theory of genes," seeking throughout his life to supplement that theory with a more dynamic and process-oriented "epigenetic theory."[8] Finally, Waddington had a bent for theoretical speculation that was conspicuously out of tune with the more strictly empirical disposition of his contemporaries.[9]

Nonetheless, some of the concerns that motivated Waddington did eventually take root, and the literature shows that by the early 1960s, talk of gene action had largely given way to a new discourse of gene activation: genes do not act all the time, but instead need to be turned on and off in response to specific stimuli. But the primary route by which the concept of differential gene activation entered into the history of molecular genetics had little to do with Waddington; and as the story is usually told, even Beermann's evidence played only a secondary role. For the majority of geneticists of the time (particularly in the Anglo-American world), the impetus for this new interest in gene activation came not from either cytological or genetic studies of biological development but from biochemical and genetic studies of bacterial adaptation. And here, the work of François Jacob and Jacques Monod, already discussed in Chapter 2, was of paramount importance. Their work not only demonstrated the fact of gene regulation but also described the mechanisms involved in such regulation.

By any measure, Jacob and Monod's analysis of gene regulation must count as one of the major triumphs of early molecular biology, but perhaps even more influential was their description of how such regulation is achieved. By calling these mechanisms "genetic regulatory mechanisms" and not "mechanisms of gene regulation," they implied that such mechanisms are themselves genetic, laying to rest any notion that genes might rely on nongenetic factors for instructions as to when and where to act. In their paper in the *Journal of Molecular Biology* Jacob and Monod concluded:

"The discovery of regulator and operator genes . . . reveals that the genome contains not only a series of blue-prints, but a coordinated program of protein synthesis and the means of controlling its execution."[10]

For Waddington (as for Beermann), differential gene activation posed a potential challenge to the autonomy of genes, and possibly even to their primacy as causal agents. In Jacob and Monod's view, by contrast, genes may need to be activated, but other genes—regulator genes—were there to do the job. The net effect of Jacob and Monod's description of a gene-based mechanism of regulation was to put genes back in the driver's seat and traditional expectations of genetic control safely back on track.

## The Genetic Program

Jacob and Monod's use of the term *program* in this paper was, to my knowledge, the first occurrence of this word in the literature of molecular biology, and it caught on rapidly.[11] It introduced a new metaphor for thinking about development, one with distinct advantages over the earlier notion of gene action. The metaphor of a program allowed for gene interactions in ways that the earlier metaphor did not; it resonated powerfully with recent developments in computer science; and, most important of all, it could encompass the new work on gene regulation. But Jacob and Monod's innovation was not simply the proposal of a program for development; it was their proposal of a program entirely contained within the genome. It was, in other words, the notion of a "genetic program."

In his immensely popular recounting of the history of

heredity published a few years later, Jacob described the organism as "the realization of a programme prescribed by its heredity."[12] Furthermore, he argued, "when heredity is described as a coded programme in a sequence of chemical radicals, the paradox [of development] disappears."[13] Jacob saw the genetic program, written in the alphabet of nucleotides, as the source of the apparent purposiveness of biological development. Referring to the oft-quoted characterization of teleology (the doctrine of final causes in nature) as a "mistress" whom biologists "could not do without, but did not care to be seen with in public," he wrote, "The concept of programme has made an honest woman of teleology."[14] Jacob felt no need to define the term *program*; he simply observed that it "is a model borrowed from electronic computers. It equates the genetic material of an egg with the magnetic tape of a computer."[15]

Without question, computers have provided an invaluable source of metaphors for molecular biology, the metaphor of a program being only one of many. But computers cannot take sole credit for the notion of a genetic program. Compelling as the analogy may be, equating the genetic material of an egg with the magnetic tape of a computer does not imply that that material encodes a program; it might, for example, just as well be thought of as encoding data to be processed by a program located elsewhere in the cell.[16] Indeed, over the very same decade in which the genetic program grew to such popularity among molecular biologists, another quite different use of the program metaphor was being exploited by computer scientists and developmental biologists—the notion of a developmental program. In contrast to the genetic program, the developmental program was not located at particular sites (for example, the genome) but was assumed, instead, to be distributed throughout the fertilized egg.

To take but one example, in 1965 a young graduate student trained in information theory and cybernetics, Michael Apter, teamed up with the developmental biologist Lewis Wolpert to argue for a direct analogy not between computer programs and the genome but between computer programs and the (fertilized) egg: "If the genes are analogous with the sub-routine, by specifying how particular proteins are to be made . . . then the cytoplasm might be analogous to the main programme specifying the nature and sequence of operations, combined with the numbers specifying the particular form in which these events are to manifest themselves . . . In this kind of system, instructions do not exist at particular localized sites, but the system acts as a dynamic whole."[17] By the mid 1970s, however, even Wolpert had been converted to the notion of a genetic program.[18]

What was the evidence that persuaded him, along with so many others? Perhaps more to the point we might ask: What exactly *is* a genetic program? James Bonner, an expert on the biochemistry and physiology of regulation in plants, put the problem well: "Of what does the programme consist and where does it live?"[19] His 1965 book, *The Molecular Biology of Development*, was devoted to answering this question. In many ways, Bonner's efforts now look outdated, but they are instructive nonetheless. Of particular interest is the ease and rapidity with which he answers the second part of his question—where the program resides—for in that very ease and rapidity, Bonner provides us with key insights into how the assumption of a program located in the genome came so quickly, and so widely, to prevail. It is worth pausing, therefore, to try to unpack his reasoning.

## THE LOGIC BEHIND THE "GENETIC PROGRAM"

Bonner's starting point was the recognition that the central dogma of molecular biology gives us only a "description of the manner in which all cells are similar." It leaves unexplained how cells of higher organisms come to be different.[20] As he wrote: "Each kind of specialized cell of the higher organism contains its characteristic enzymes but each produces only a portion of all the enzymes for which its genomal DNA contains information." Immediately, however, he continues with the rudiments of an answer in view: "Clearly then, the nucleus contains some further mechanism which determines in which cells and at which times during development each gene is to be active and produce its characteristic messenger RNA, and in which cells each gene is to be inactive, to be repressed."[21]

Two important moves have been made here. Bonner's main point is to argue that something other than the information encoded in the DNA for protein synthesis is required to explain cell differentiation, but on the way to making this point he has placed this "further mechanism" in the nucleus, with nothing more by way of argument or evidence than his "Clearly then." Why does such an inference follow? And why does it follow "clearly"? The next paragraph offers some help: "The egg is activated by fertilization . . . As division proceeds, cells begin to differ from one another and to acquire the characteristics of specialized cells of the adult creature. There is then within the nucleus some kind of programme which determines the properly sequenced repression and derepression of genes and which brings about orderly development."[22]

Here, Bonner explicitly refers to the required "further

mechanism" as a program, and once again he has located it in the nucleus. But this time around, a clue to the reasoning behind his inference has been provided in the very first sentence, "The egg is activated by fertilization." This is how I believe the (largely tacit) reasoning goes: If the egg is "activated by fertilization," the implication is that it is entirely inactive prior to fertilization. What does fertilization provide? The entrance of the sperm, of course. Unlike the egg, the sperm contains almost no cytoplasm and accordingly can be thought of (and has indeed been so figured throughout its history) as pure nucleus. Ergo, the active component must reside in the nucleus and not in the cytoplasm.

Aeschylus, in *The Eumenides,* put just such a view into the voice of Apollo: "I will tell you, and I will answer correctly. Watch. The mother is no parent of that which is called her child, but only nurse of the new-planted seed that grows. The parent is he who mounts. A stranger she preserves a stranger's seed, if no god interfere." Of course, Mendel's laws had long since put a decisive divide between modern views of generation and those of Aeschylus' time, and in recent years it might be said that the divide has grown greater still. In all likelihood, the supposition of an inactive cytoplasm would be challenged today. But in 1965 it would have been taken for granted—indeed, as a yet-to-be questioned residue of the classical discourse of gene action, it would almost surely have gone unnoticed. As if by way of corroboration of this point, Bonner goes on to add: "We can say that the programme which sequences gene activity must itself be a part of the genetic information since the course of development and the final form are heritable. Further than this we cannot go by classical approaches to differentiation."[23]

In the space of less than two paragraphs, Bonner has now completed the line of argument that leads him to his conclusion: the program *must* be part of the genetic information. And once again, we can try to unpack his reasoning: Why does the heritability of "the course of development and the final form" imply that the program must be part of the genetic information? Because of the unspoken assumption that since the course of development and the final form are heritable, the only causally relevant material to be transmitted from one generation to another is the genetic material. The manifest fact that the reproductive process passes on not only genes but also cytoplasm is not mentioned. Nor need it be. Given the prior assumption that the cytoplasm contains no active components, this fact would almost certainly have been regarded as irrelevant. Indeed, the conviction that the cytoplasm could neither carry nor transmit effective traces of intergenerational memory had been a mainstay of genetics for so long that it had become part of the "memory" of that discipline, working silently but effectively to shape the very logic of inference.

The remainder of Bonner's book is devoted to answering the first part of his question, "Of what does the programme consist?" and in the final chapter, he attempts to sketch out an actual computer program for development. Here Bonner seeks to reframe what is known about the induction of developmental pathways during the life cycle in terms of a "master programme constituted in turn of a set of subprogrammes or subroutines."[24] Each subroutine specifies a specific task to be performed. For a plant, his list includes cell life, embryonic development, how to be a seed, bud development, leaf development, stem development, root development, reproductive development, and others. Within each of these subroutines is a list of cellular instructions or commands. For example, "Divide tangentially with growth"; "Divide transversely with growth"; "Grow without dividing"; "Test for size or cell number."[25] He then asks, "How might these subroutines be related to one another? Exactly how are they to be wired together to constitute a whole programme?"[26] Bonner's answer comes in the form of a provisional map of the "switching network" which he sometimes describes as "developmental" but more frequently as "genetic." And here, in his discussion of such switching networks, two final points need to be underscored, for they will bear directly on my later discussion of genetic programs.

The first has to do with the structure Bonner proposes for the required networks. Although his subroutines are laid out in a list or linear sequence—as if following from an initial master program—in fact they constitute a circle, as indeed they might be expected to if they are to describe a life cycle.[27] The second has to do with how the instructions specified in the various subroutines comprising the life cycle are actually embodied. Given the dependence of development on the systematic and regulated activation of particular genes, and given Jacob and Monod's clear demonstration that particular genes need to be switched on and off, Bonner's description of the switching network for developmental processes as a "genetic switching network" may seem reasonable enough. But the phrase harbors a potentially treacherous ambiguity.

Does the word *genetic* refer to the subject of the switching network or to its object? Does it refer to the regulator or to that which is regulated? In the words of a recent article in *Science*, "Whose finger is on the switch?"[28] The reading that Bonner clearly intends is that the network is both constituted of and controlled by genes. Ironically, however, closer scrutiny of the provisional map that he himself proposes for such a network reveals that many other kinds of entities figure as well, and all of these play critical roles in the control of genetic activity.[29] The very ambiguity of the term *genetic switching network* thus invites what amounts to a category error. Perhaps it does not need saying, but this is precisely the ambiguity that plagues the term *genetic program*. Does the word *genetic* refer to the subject or to the object of the program? Are the genes the source of the program, or that upon which the program acts?

## Learning How to "Reprogram" the Genomes of Higher Organisms

To say that a great deal has happened since those early years would once again be an understatement. This time around, however, the relevant developments have been not only in molecular biology but in reproductive biology and computer science as well. We still speak of programs, but the meaning of that term—in biology and in computer science—has changed significantly. In both fields, programs have come to be understood as multilayered and distributed. To be sure, the informational content of DNA remains essential—without it, development (life itself) cannot proceed. But current research in a number of biological disciplines has begun to put considerable pressure on biologists to reconceptualize the program for development as something considerably more complex than a set of instructions written in the "alphabet of nucleotides" and more nearly resembling earlier notions of a developmental program.

As before, much of the most persuasive evidence has come from new techniques in molecular biology, and perhaps especially from interventions enabled by new sequence data. But sequence information is not the only route by which biologists have come to shift their conceptual focus from a genetic program to a developmental program. In fact, in the one area where this shift may be most evident, sequence information has yet to play much of a role at all. I am referring to work over the last twenty-five years on "cloning" by "nuclear transfer," and especially to dramatic successes in growing new organisms from zygotes that have been formed by fusing an adult cell of one animal with, or transferring the nucleus of that cell into, an enucleated oöcyte (or oöplast) of another of the same species.[30]

Dolly the sheep was the first mammal to be cloned in this way, and the news of her birth brought instant acclaim to the scientists who had achieved this remarkable feat.[31] But why was it so remarkable? In fact, what made it such a surprise? If the program for development resides in the genome, why shouldn't it be possible to clone a new organism from the nucleus of an adult cell? Why, for that matter, should one need to transfer the nucleus into an oöplast? What's wrong with the cytoplasm of an ordinary cell?

The central problem in cloning new organisms from adult cells is that adult cells are specialized, and their reproduction normally leads only to more cells of the same kind. Indeed, this fact by itself ought to give us pause. If all cells have the same DNA, why should fully differentiated adult cells reproduce only their own kind? Where do the instructions governing the expression of a particular cell type reside? And how are these instructions passed on from one generation of cells to another? If cell differentiation is not accompanied by changes in the DNA itself, then the clear implication is (1) that other, *epigenetic*, factors (possibly residing in the cytoplasm) must be involved in determining cell type, and (2) that the stability of cell type requires mechanisms of epigenetic inheritance.[32]

It has been recognized for more than half of a century that changing the cellular environment of the nucleus of a cell could alter the fate of the daughter cells. Yet no amount of manipulation of the cytoplasm proved effective in stimulating the growth of a new organism. Thus, the assumption was that the nuclei of adult cells had lost their totipotency— that is, that once they were programmed to produce a particular kind of cell, and as a consequence of this programming, they no longer had the capacity to generate all the different kinds of cells that are needed for a new organism. In fact, until it was demonstrated by nuclear transfer that the nuclei of differentiated cells are capable of guiding and directing normal embryonic development, biologists could not be absolutely certain that such programming did not lead to irrevocable changes in the DNA of differentiated cells.

The first success in inducing embryonic development in higher organisms by nuclear transplantation was achieved by Robert Briggs and T. J. King in 1952.[33] But it was largely as a result of John Gurdon's work in the 1970s that biologists learned of the possibility, in frogs at least, of "reprogramming" the nucleus of a fully differentiated cell by transferring it into an enucleated zygote—in some cases, well enough to support complete embryonic development.[34] Mammalian development seemed to pose special problems, however. Despite repeated attempts, only partial reprogramming of transferred nuclei could be achieved, and the genomes of mammalian cells were accordingly described as having "inertia": they "resisted" being reset in their original functional state. This, then, is this background against which Ian Wilmut and his colleagues were working—and the reason for the great surprise that greeted their announcement of Dolly's birth in 1997.[35] With that achievement, they demonstrated that the mammalian genome does not undergo irreversible change in the course of development (as some had suspected) but rather that the obstacle to reprogramming lay in the relationship that had been established between nucleus and cytoplasm. As Colin Stewart wrote, "The key to success seems to have been in finding a method to make the donor nuclei more compatible with the cytoplasm of the recipient oocyte."[36]

The key word here is *method*. To this day, little is known about the molecular basis of this compatibility, and success in finding an effective method depended considerably more on tricks of the trade—with a large admixture of trial and error—than on an understanding of what reprogramming actually involves, or even of what the term precisely means. It is notable, for example, that the object of the verb "to reprogram" in this literature is almost always the nucleus and only rarely the genome. Is there a difference? In fact, there is, and the difference is almost certain to be important.

It has been known for many decades that genomes of eukaryotic organisms never appear in their naked state. Rather, they come in tightly bound complexes of DNA and chromosomal proteins known as *chromatin* (see Figures 8 and 9).[37] The specific structure of the chromatin is now recognized as playing a critical role in gene expression and might be said to constitute the most immediate context of the genetic material. Thus, a prime task in reprogramming

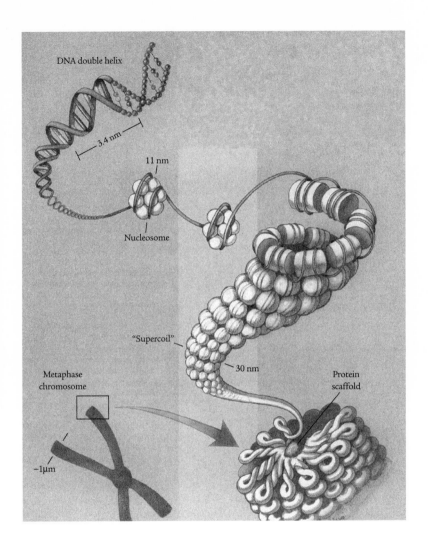

DNA double helix

3.4 nm

11 nm

Nucleosome

"Supercoil"

30 nm

Metaphase
chromosome

Protein
scaffold

~1μm

Figure 8: DNA packaging: Watson and Crick established the famous double-helix structure of DNA (top left). More recent studies have revealed the structure of a higher-order of packaging (center right), of the about 2 meters of DNA and nucleosome cores in eukaryotic cells, before it folds into chromosomes (bottom left). These structures play a crucial role in gene regulation, DNA repair, and reproduction. *(Reproduced by permission of the U.S. Department of Energy.)*

## Levels of Chromatin Packing

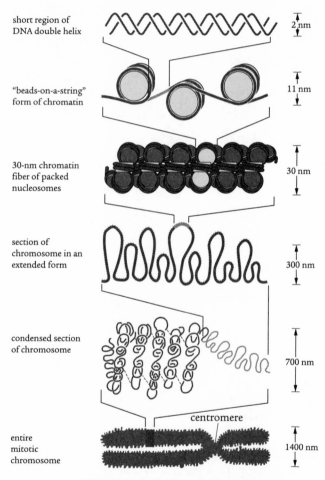

short region of
DNA double helix

2 nm

"beads-on-a-string"
form of chromatin

11 nm

30-nm chromatin
fiber of packed
nucleosomes

30 nm

section of
chromosome in an
extended form

300 nm

condensed section
of chromosome

700 nm

centromere

entire
mitotic
chromosome

1400 nm

NET RESULT: EACH DNA MOLECULE HAS BEEN
PACKAGED INTO A MITOTIC CHROMOSOME THAT
IS 50,000x SHORTER THAN ITS EXTENDED LENGTH

**Figure 9: An artist's rendering of DNA packaging.** A different view of the
higher-order structure of DNA depicted in Figure 8. *(From B. Alberts et al.,*
Essential Cell Biology, *1998; reproduced by permission of Garland Publishing.)*

the nucleus is remodeling and restructuring the chromatin, for therein lie the agents most immediately responsible for regulating gene expression.

Yet here as well, to speak of agency is somewhat deceptive, because the secret of successful cloning lies in the receptivity of the chromatin to remodeling and restructuring by more distal cues from the cytoplasm. The trick that Wilmut and his colleagues employed was to place the donor cells in a quiescent state, for, as Keith Campbell writes, "The chromatin of the quiescent cell may be more amenable to structural changes after nuclear transfer which are associated with 'reprogramming' of gene expression."[38] But as to what makes the chromatin of quiescent cells more amenable, or what kinds of cytoplasmic cues are responsible for inducing such changes, even now very little is known.

Writing more than two years after their original triumph, Campbell concludes: "At the present time the only true measure of 'reprogramming' is the production of live offspring. The mechanisms underlying the spatial and temporal control of gene expression, imprinting, X chromosome inactivation are complex. Hopefully the technique of nuclear transfer will help to elucidate some of the underlying mechanisms."[39]

## THE VIEW FROM MOLECULAR GENETICS TODAY

Forty years ago, with the publication of Jacob and Monod's elegant model for gene regulation, many believed that all biological development might be understood in terms of the operon model and that the problem of development—at least in principle—had therefore been solved. But writing thirty years later, Sydney Brenner referred to the view that development "was simply a matter of turning on the right genes in the right places at the right times" in rather scathing terms: "Of course, while absolutely true this is also absolutely vacuous. The paradigm does not tell us how to make a mouse but only how to make a switch. The real answer must surely be in the detail."[40] But Brenner had written this years before we learned how to clone mice. In any case, his judgment would have had little to do with the efforts of those who were developing techniques of nuclear transfer in mammals. Brenner is a molecular geneticist. He is widely known not only as one of the major figures of the original molecular revolution (and a close colleague of Monod and Jacob's) but also as a pioneer of more recent developments in molecular genetics. And surely, no one knows better than Brenner that the signature of molecular genetics is precisely its capacity to produce such detail. Today, as we peruse the evidence that has accumulated over the last two decades, there can be little doubt that Brenner was right: the secret to making an organism *is* in the details.

Even so, beliefs in a centralized program for development that resides in the genome have not died, and especially not in the popular press. They endure most conspicuously in the continuing use of the very term *genetic program.* When, for example, the complete sequence of *C. elegans* (the roundworm that Brenner himself had established as a model organism) was announced on December 11, 1998, it made big news. *The New York Times* carried the story on its front page under the title, "Animal's Genetic Program Decoded, in a Science First," and the first sentence read: "Biologists have for the first time deciphered the full genetic programming of an animal."[41] Only six paragraphs down do we read of the other side of the story: "Seeing the worm's complete genome is humbling, Dr. Alberts [president of the National Academy of Sciences] said, because it makes biologists realize how much there is yet to understand. 'We always underestimate the complexity of life, even of the simplest processes,' he said. 'So this is really only the beginning of unraveling the mystery of life.'"[42]

To say this is not to underestimate the importance of having the full sequence of the genome. It is only to emphasize the extent to which that importance resides in the use of sequence data as a tool, as a way of probing the complexity of developmental dynamics. Indeed, it is precisely such data that have enabled molecular biologists to recognize the dense entanglement of developmental processes and, in that recognition, to appreciate the limits of centralized control. As Antonio Garcia-Bellido writes, "Development results from local effects, and there is no brain or mysterious entity governing the whole: there are local computations and they explain the specificity of something that is historically defined."[43]

To be sure, not all molecular geneticists see things this way. Indeed, there is a noticeable tension in the contemporary literature on the utility of a notion of centralized control. For example, Eric Davidson and his colleagues have conducted an elegant and exhaustive analysis of the functional properties of the promotor region of a gene playing a particularly crucial role in the development of sea urchins. On the basis of their analysis, they have constructed a computational model "that explicitly reveals the logical interrelations hard-wired into the DNA."[44] "Perhaps the main insight from this experimental exploration," they conclude, "is that these system properties are all explicitly specified in the genomic DNA sequence."[45] Without question, their model constitutes a major breakthrough in our understanding of developmental regulation; yet the actual details are somewhat at odds with this description. While the DNA does indeed provide the original sequences (what computer scientists might call the source code) used in the construction of the many proteins participating in these interactions, the relevant sequences are scattered throughout the genome. Furthermore, the dynamics of interaction between proteins and DNA binding sites (whether, for example, a protein functions as an activator or inhibitor) are often determined by features of protein structure that are themselves subject to cellular regulation.[46]

In a similar vein, Walter Gehring and his colleagues made headlines when they announced that targeted activation of the *eyeless* gene in *Drosophila* induces the formation of full-fledged eyes in fly wings, legs, antennae, and a variety of other tissues that don't normally produce eyes. Because of this success, Gehring suggested that *eyeless* might be "the master control gene for eye morphogenesis."[47] One year later, his proposal seemed to be corroborated by the finding that a homologous gene found in the mouse (that is, a gene with extensive regions matching the sequence of *eyeless*) would, when inserted into *Drosophila,* work the same magic as *eyeless.* One might however ask, exactly what does this finding demonstrate? As Gehring himself writes, "Of course, these eyes were *Drosophila* eyes . . . because the mouse provided only the switch gene and *Drosophila* contributed the other 2,500 genes required to make an eye."[48]

The eye has a rather special significance in the history of science. At least since Darwin's day, it has served as the exemplar of design—as a prototype of those "organs of ex-treme perfection and complication" the formation of which has for so long defied scientific explanation. As Darwin him-self acknowledged, "To suppose that the eye, with all its inimitable contrivances . . . could have formed by natural se-lection, seems, I freely confess, absurd in the highest possi-ble degree."[49] Given this background, discovery of "the mas-ter control gene" responsible for eye morphogenesis was news indeed. But there is a sense in which this claim is obvi-ously contradicted by the very experiment that has been taken to corroborate it. If the mouse counterpart to *eyeless* (*Pax*-6) were truly a "master control gene," ought we not to expect that it would induce the formation of a mouse eye and not a *Drosophila* eye? Might one not interpret the fact that the mouse gene is used by the fly to form its own kind of eye as corroborating a claim of a rather different sort— namely, that *eyeless* plays a key role in the formation of an eye, the precise nature of which is determined by the context in which the gene finds itself?

In much the same spirit, an alternative reading of this work is prompted by asking, as we did in Chapter 2, what does *eyeless* do? A member of the class of genes known as *homeotic* genes, *eyeless* is said to encode a key transcription factor (a protein that regulates the expression of a num-ber of other genes by its specific capacity to bind to par-ticular sequences of the DNA). But as with so many other eukaryotic genes, the sequence encoding the relevant pro-tein must be constructed post-transcriptionally by splicing. In this sense, the DNA sequence of *eyeless* might better be re-garded as a potential gene. In his recent book, Enrico Coen suggests an alternative description—not "master genes" but "master proteins" (see Figure 10). Coen writes, "You can think of the binding site [on the DNA] as a sort of lock, and the matching protein as the key that fits it."[50]

But as Coen clearly recognizes, "master genes" and "master proteins" are both inadequate descriptions of developmental dynamics, and perhaps equally so. They are reminiscent of the age-old conundrum of which came first—the chicken or the egg. What makes that question a conundrum is that the answer is so obviously "Both." Indeed, the fact that it is not possible to have one without the other may be taken as the defining feature of an organism. As Coen puts it, "Organisms, from daisies to humans, are naturally en-

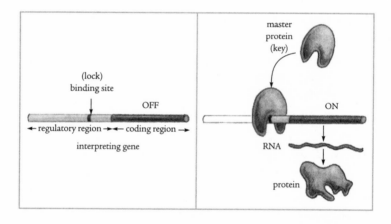

Figure 10: **Master proteins.** Structure of an interpreting gene showing the regulatory region, which contains a binding site, and the coding region, which is used to make RNA (which in turn leads to protein being made). In the left panel, no master protein is present and the interpreting gene is *off.* In the right panel, a master protein is present and it binds to the regulatory region: this encourages the coding region to be transcribed into RNA so the gene is *on.* (*From E. Coen,* The Art of Genes: How Organisms Make Themselves, *1999; reproduced by permission of Oxford University Press.*)

dowed with a remarkable property, an ability to *make them-selves*."[51] Hence the title of his book: *The Art of Genes: How Organisms Make Themselves*. Coen's starting point is the "revolution in biology" that has occurred over the past twenty years, and it is a revolution he has both participated in and observed.

Virtually everyone working in developmental genetics today would agree with these words, even if they might differ about the best way to articulate this new understanding. Coen would like to dispense with the very notion of a program, but it may not be necessary to give up on the term altogether. If we wanted to keep the computer metaphor, we could describe the fertilized egg as a massively parallel and multilayered processor in which both programs (or networks) and data are distributed throughout the cell.[52] The roles of data and program here are relative, for what counts as data for one program is often the output of a second program, and the output of the first is data for yet another program, or even for the very program that provided its own initial data. For some developmental stages, the DNA might be seen as encoding programs or switches that process the data provided by gradients of transcription activators. Or, alternatively, one might say that DNA sequences provide data for the machinery of transcription activation (some of which is acquired directly from the cytoplasm of the egg). In later developmental stages, the products of transcription serve as data for splicing machines, translation machines, and so on. In turn, the output from these processes make up the very machinery or programs needed to process the data in the first place.

One of the most important discoveries of the 1980s was that close homologues of the genes required for building the basic body plan of an animal—the homeotic genes—could be found across the animal kingdom. In the 1990s we learned that chimps share 98.5 percent of human DNA. Indeed, Peter Holland writes, "We can now state with confidence that most animal phyla possess essentially the same genes."[53] Nevertheless, the differences between flies and mice, and even between chimps and humans, are unmistakable. If the genes are "essentially the same," what then is it that makes one organism a fly and another a mouse, a chimp, or a human? The answer, it seems, is to be found in the structure of gene networks—in the ways in which genes are connected to other genes by the complex regulatory mechanisms that, in their interactions, determine when and where a particular gene will be expressed. But unlike the sequence of the genome, this regulatory circuitry is not fixed: it is dynamic rather than static, a structure that is itself changing over the course of the developmental cycle. Indeed, it is just this dynamic system that I am calling the developmental program. Exactly how its circuitry changes is of course key to the character of the end result of development. But along with an increasing number of others, I argue that an understanding of its dynamics needs to be sought at least as much in the interactions of its many components as in the structure or behavior of the components themselves.[54]

In brief, then, my take on the revolution to which Coen refers is that it is a revolution still in the making. And I suggest that we might get a measure of how far it has come by revisiting Bonner's question from 1965. Today, if we were to ask, "Of what does the program consist and where does it live?" we would hear a growing number of researchers in the field telling us that the program consists of, and lives in, the interactive complex made up of genomic structures and the vast network of cellular machinery in which those structures are embedded. It may even be that this program is irreducible—in the sense, that is, that nothing less complex than the organism itself is able to do the job.

## 4

## Limits of Genetic Analysis:
## What Keeps Development on Track?

The complex accomplishment of any one living cell is part and parcel of the [fact] that any one cell represents more an historical than a physical event. These complex things do not rise every day by spontaneous generation from the non-living matter—if they did, they would really be reproducible and timeless phenomena, comparable to the crystallization of a solution, and would belong to the subject matter of physics proper. No, any living cell carries with it the experiences of a billion years of experimentation by its ancestors. You cannot expect to explain so wise an old bird in a few simple words.

Max Delbrück, "A Physicist Looks at Biology" (1949)

Addressing the thousandth meeting of the Connecticut Academy of Arts and Sciences in 1949, Max Delbrück reflected back upon his experience as a physicist who had turned his attention to biology just a little more than a decade before. And here, in his opening words, he put his finger on the single most important distinction between the two sciences. "A mature physicist, acquainting himself for the first time with problems of biology, is puzzled by the circumstance that there are no 'absolute phenomena' in biology."[1] Biology is not lawful in the same sense in which physics is, for every feature of a biological organism is what it is by virtue of its long evolutionary history. And the reason the outcome of all these "billion years of experimentation by its ancestors" is never either absolute or predictable is that the experimental materials with which primitive life forms could work were themselves dependent on the occurrence of chance events. Life as we know it is the beneficiary of this long history of fortuitous opportunities. Stephen Jay Gould likens evolution to a videotape that, if replayed over and over, would have a different ending with each playing.[2] In fact, it is sometimes argued that chance, or contingency, is the defining characteristic of evolution, and possibly even its driving force.

But if contingency is the key to evolution, it might be argued that the obverse of contingency—the capacity to stay on track despite the myriad vicissitudes that inevitably plague a developing organism—is the key to biological development. Over the course of its development, the nascent organism must withstand not only the relentless variability of its immediate external environment but, equally, the local fluctuations in its internal environment. To underscore the difference between evolution and development we might borrow Gould's metaphor and liken development to a videotape that displays countless variations of plot each time it is played, yet always concludes with essentially the same ending.[3]

In fact, the very stability—or robustness—of the developmental process might be said to be a precondition for evolution by natural selection. Selection, as we know, acts on phenotype, not on genotype, and without the capacity to develop normally despite the vicissitudes of internal and external environment, organisms of a particular genotype would not develop unto a particular phenotype with sufficient reliability for selection to act upon them.[4] Because a given phenotype can often be produced by a variety of different developmental and molecular pathways, phenotypic stability might even be said to *exceed* genotypic stability. Occurrences of this kind reflect the capacity of developmental dynamics to resist not only intra- and intercellular variation but also some degree of genetic variation of a kind that is compensated for in the course of development and therefore not normally expressed.[5] Furthermore, even when variations in individual traits do occur, in each generation the fertilized egg of a particular species grows, with astonishing dependability, into an adult that is still clearly recognizable as a member of that species.

Over the last two decades, we have learned an immense amount about the elaborate mechanisms involved in producing an adult phenotype from a fertilized egg, and the delicacy and precision of these mechanisms might well fill us with awe. Yet that very delicacy and precision present a problem, just because they depend on the exact positioning of what is often only a very small number of molecules. Schroedinger wondered over the "remarkable" capacity of an organism to withstand the forces of disorder, but perhaps his attention was misplaced. I would argue that even more remarkable than the persistence of the material gene structure through so many generations is the reliability with which an individual organism, in each generation, negotiates its precarious passage from zygote to adult. How, we might ask, is such impressive reliability ensured? How does a developing organism manage such success in reaching its final goal? Questions of this sort may have the ring of an era long past—indeed, they were once the central questions of biology. But over the last decade they have reemerged with surprising insistence. And with their reemergence, biologists have acquired new appreciation of the appropriateness of Max Delbrück's observation that "You cannot expect to explain so wise an old bird in a few simple words."

## WHAT IS AN ORGANISM?

Loosely understood, the question "What is an organism?" may be taken as marking the beginnings of biology as a separate and distinctive science. The field's emergence is commonly linked to the coining of the word "biology" in 1802—as it happens, independently by three authors. For each of these authors, the new term demarcated the world of the living from that of the nonliving and, with that demarcation, named a new "Science of Life."[6]

But what was it that so clearly distinguished the living from the nonliving for these scientists at the end of the eighteenth century? Their answer was contained in a single word: *organization*. What made it possible to distinguish an organism from its Greek root *organon* (or tool) was the special arrangement and interaction of parts that bring the well-springs of form and behavior *inside* the organism itself. A tool, of necessity, requires a tool-user, whereas an organism is a system of organs (or tools) that behaves as if it had a mind of its own—as if it governs itself.

Indeed, the two words *organism* and *organization* acquired their contemporary usage more or less contemporaneously. Immanuel Kant, in 1790, gave one of the first modern definitions of an organism—not as a definition per se but rather as a principle or maxim which, he wrote, "serves to define what is meant as an organism"—namely, "*an organized natural product is one in which every part is reciprocally both end and means.* In such a product nothing is in vain, without an end, or to be ascribed to a blind mechanism of nature."[7] Organisms, he continued, are the beings that "first afford objective reality to the conception of an *end* that is an end of *nature* and not a practical end. They supply natural science with the basis for a teleology . . . that would otherwise be absolutely unjustifiable to introduce into that science—seeing that we are quite unable to perceive *a priori* the possibility of such a kind of causality."[8] Elaborating upon this kind of causality, he wrote: "In such a natural product as this every part is thought as *owing* its presence to the agency of all the remaining parts, and also as existing *for the sake of the others* and of the whole, that is, as an instrument, or organ . . . The part must be an organ *producing* the other parts—each, consequently, reciprocally producing the others . . . Only under these conditions and upon these terms can such a product be an *organized* and *self-organized being,* and, as such, be called a *physical end.*"[9] Indeed, it is here that the term *self-organized* first makes its appearance in relation to living beings. It is invoked—and underscored—to denote Kant's explicit opposition to argument by design. No external force, no divine architect, is responsible for the organization of nature, only the internal dynamics of the being itself.

Thus the beginnings of biology prescribed not only the subject and primary question of the new science but also the form of answer to be sought. To say what an organism is would be to describe and delineate the particular character of the organization that defined its inner purposiveness, that gave it a mind of its own, that enabled it to organize itself. What is an organism? It is a bounded, physicochemical body capable not only of self-regulation—self-steering—but also, and perhaps most important, of self-formation. An organism is a material entity that is transformed into an autonomous and self-generating "self" by virtue of its peculiar and particular organization.

Fascination with the construction of ever more elaborate automata was at a high point at the close of the eighteenth century and the dawn of the nineteenth, and craftsmen and engineers displayed extraordinary virtuosity in the building of these lifelike machines. Even so, no one was fooled. And to Kant, as to his contemporaries, it seemed evident that neither blind chance nor mere mechanism, and certainly no machine that was then available, could recreate the self-generating and self-organizing properties that were so manifest in actual living beings. "Strictly speaking," he wrote, "the organization of nature has nothing analogous to any causality known to us."[10] Thus the obvious task for biology in the years ahead was to try to understand the character of this special kind of organization, or self-organization.

Unfortunately, notwithstanding the impressive advances of nineteenth-century biological science in unraveling the physicochemical basis of such physiological processes as respiration and metabolism, little if any progress could be reported on the problem of organization. And toward the close of that century, even Claude Bernard—the man who had contributed so much to our understanding of the chemistry of physiology—seemed ready to despair. In 1878 Bernard wrote, "There is a kind of pre-established design for each being and each organ, so that, considered in isolation, each phenomenon of the harmonious arrangement depends on the general forces of nature, but taken in relationship with the others, it reveals a special bond: some invisible guide seems to direct [the phenomenon] along the path it follows, leading it to the place it occupies."[11]

But the twentieth century proved notably more beneficent. With the advent of molecular biology, biologists seemed finally to have found their homunculus, and it turned out to be, after all, a molecule. Rereading Bernard's words of 1878 with the benefit of less than one hundred years of hindsight, François Jacob could write, "There is not a word that needs be changed in these lines today: they contain nothing which modern biology cannot endorse." Jacob's solution of this age-old impasse was, as we saw in Chapter 3, the genetic program. Here, buried deep inside the innermost core of cellular structure, inscribed "in a sequence of chemical radicals," was the "invisible guide" required to direct the organism "along the path it follows, leading it to the place it occupies."[12]

Kant had been unable to envision a mechanical device capable of directing the development of an organism to its final goal for the obvious reason that, in his own time, no such automaton, no machine with comparable capacities, was there to be seen. Nor would it be for another 150 years. By the middle of the twentieth century, however, Jacob had only to look around him to see the development of a new kind of machine, a kind of mechanical device that, in one way or another, promised to bridge the gap between organisms and the machines of yesterday.

"Out of the wickedness of war," wrote Warren Weaver in 1949, in a paper entitled "Problems of Organized Complexity," "have come two new developments . . . of major importance in helping science to solve these complex twentieth-century problems."[13] The first of these was the electronic computer, built to process the masses of data generated by the procedures of modern warfare—and, perhaps most famously, to decipher enemy messages encoded in ever more elaborate encryption devices.[14] The second development is most commonly associated with the word *cybernetics,* Norbert Wiener's term for the study of control and communication in machines and living beings. Extrapolating from his experience with "goal-oriented" and "self-steering" devices designed to improve the accuracy of anti-aircraft artillery, Wiener and his followers envisioned the construction of purposive machines that would resemble living organisms in every way. Indeed, these machines would be built on the very principles of circular causality ("in which every part is reciprocally both end and means") that Kant himself had invoked as the defining feature of the organism.

These two developments were clearly related—at the very least, they were related in time, in place, and in the needs from which they arose. Yet despite their persistent conjoining in the popular imagination, despite Wiener's own hopes, and despite even John von Neumann's efforts at integration, conspicuous differences between the two remained. In the one the emphasis was on computational power, while in the other it was on principles of organization and—increasingly over the 1950s and 1960s—of self-organization.[15] In fact, in was not until the 1980s that the different visions embodied in these two developments would begin to resolve, and the first steps of that resolution came with the rise of connectionism, parallel processors, and neural networks. Yet Jacob's claim that the sequence of DNA could serve as Bernard's "invisible guide" depended absolutely on joining together these two still-disparate developments. His metaphor of a program drew directly from Turing's original model of a computer (the reader may recall from Chapter 1 his equation of "the genetic material with the magnetic tape of a computer"[16]), but the idea of a purposive machine was borrowed from Wiener's cybernetic vision. The difficulty is that, in locating the program in the genome, much of the cybernetic vision of goal-seeking and self-organization was lost. And so was the recognition of the importance of reliability and with it, an appreciation of the kinds of organizing principles that would be needed to maintain such reliability. Redundancy, for example, is a basic principle of design for building reliable systems, and it is hard to imagine how, were it not for this amnesia, recent findings of extensive redundancy in developmental pathways could have been quite as startling as they have been.

## GENES AND REDUNDANCY

The effects of gene knockouts must surely count as one of the most unsettling of the many surprises resulting from new techniques in molecular biology. Ever since the advent of gene cloning in the early 1970s, one of the primary methods used to determine the function of a gene in mammals has been to clone the gene and then search for a homologous gene in organisms in which the corresponding protein (or proteins) and its (or their) function have already been identified. A little over a decade ago, however, new techniques for targeted disruption (or knockout) of particular mammalian genes in their actual biological context have made it possible to study their function in vivo. The results surprised everyone: only rarely did such knockouts have the predicted effect. In many cases, knocking out a normal gene and replacing it with an abnormal copy had no effect at all, even when the gene was thought to be essential; indeed, in a few cases, the knockout/replacement procedure seemed to result in an *improvement* in function.[17]

In 1993, news of these surprising results reached the science section of *The New York Times:* "Rodents shorn of genes once deemed essential to life often manage just fine—evidence, scientists say, that there is extensive redundancy built into an animal's blueprint. If one important gene is deleted from an animal's DNA, other genes apparently can stand in for the missing player . . . 'It looks like different circuits are used at different times during development, and the organism has choices,' [Dr. Capecchi] said. 'If a problem is encountered, the thing has to figure out a solution. Sometimes the solution is fantastic, other times it's less so' . . . Dr. Capecchi said that given the many molecular mistakes that arise during the creation and growth of an organism, the built-in genetic redundancy is surely essential to survival. 'If we didn't have extensive overlap and redundancy in our genome,' the sum total of our genes, [he] said, 'we probably wouldn't be here at all.'"[18]

In point of fact, however, the phenomenon of redundancy was not entirely new. Evidence for extensive polymorphism in the genotypes of species living in the wild had been accruing since the 1950s.[19] And in experimental genetics, mutants exhibiting no phenotypic effect even when both copies (or alleles) of the gene in question were dysfunctional had been observed since the 1960s. Such mutants were called "null" because of their failure to produce the enzymes that were known to be produced by the wild-type (nonmutant form), and the lack of a phenotypic effect came as a considerable surprise when first identified. But it is only over the last decade that the nonappearance of expected effects has aroused serious concern.

Thanks to knockout techniques, the number of cases of such "null effects" has grown exponentially, and the consensus among researchers in the field today is that they clearly indicate the existence of widespread functional redundancy in genetic pathways. Indeed, functional redundancy—at the level of transcription, transcriptional activation, genetic pathways, and intercellular interactions—has emerged as a prominent feature of developmental organization in complex organisms, and its emergence had been generating consternation in the scientific literature for several years before hitting *The New York Times*. "Redundancy strikes fear in the heart of geneticists," wrote Sydney Brenner and his colleagues in 1990, and the reasons for fear were apparent to all.[20]

First, the invisibility of a phenotypic effect in "null" mutants reveals distinct limits to the usefulness of genetic analysis in probing developmental dynamics. The core techniques of genetics depend on the identification of mutants by their phenotypic effects (mutation screening), but the absence of a distinctive phenotypic marker makes mutants of this kind invisible to such screens. Genetics, as it is sometimes said, is blind to redundancy. But more worrisome yet, redundancy poses a threat to the entire explanatory framework of the genetic paradigm—a problem considerably more serious than a blind spot of laboratory method. Diethard Tautz writes, "Though the geneticist will often be unable to say exactly how a certain mutation causes a certain phenotype . . . he must maintain that single and direct causal relationships exist. This genetic paradigm is at the basis of all systematic mutagenesis experiments, which aim to obtain particular phenotypes, since these experiments usually allow one to look only at the effects of a single mutation at a time . . . [But] even the best paradigm eventually meets a crisis. Such a crisis is imminent."[21]

Tautz is prompted to speak of a crisis in the traditional paradigm of genetics not only because redundancy is technically opaque to the methods of genetics but also because, from an evolutionary perspective, redundancy doesn't seem to make a lot of sense. If genes are assumed to be the units of selection, how could redundant genes have evolved? Furthermore, many of the knockout genes that have been shown to display little if any phenotypic effect were genes that were initially assumed to serve an important function because of the extent to which they have been conserved over evolution. But how, in the absence of any obvious selective advantage to the organism bearing them, are we to account for their persistence? As J. H. Thomas puts it: "It is perhaps surprising that redundancy is so prevalent, since it is not immediately obvious what selective advantage it might confer. Possession of two fully redundant genes should, on evolutionary time scales, be an unstable condition . . . [A] similar argument might suggest that even partially redundant genes would tend to lose their redundancy."[22]

One of the main lessons from the early work on information theory in the 1950s was that fidelity in the transmission of information requires redundancy. Now, provoked by the need to make evolutionary sense of redundancy, Tautz recalls this lesson and goes on to suggest an obvious analogue for living systems: "The formation of an adult organism can be seen as the transmission of information which is laid down in the egg and its genome . . . At each [developmental step] there is a potential loss of information and the developing organism has to safeguard itself against this loss. This is, of course, a good basis for selection pressure to evolve redundancies. This selection pressure need not be very high, since even a small effect on the probability of successful completion of embryogenesis would directly be reflected in the probability of survival of the offspring . . . Thus, the evolution of redundant regulatory pathways may be seen as a logical consequence of the evolution of complex metazoan life."[23]

But note: in order to make use of this lesson from information theory, Tautz has been obliged to displace the gene as the unit of selection by that of the whole organism—or, more accurately, by that of the organism's life cycle. The selection pressure required for the evolution of redundancies operates not on the survival of individual genes but on the survival of the offspring! In other words, it is the endurance of the life cycle itself that has here become the subject of evolution.

LEARNING FROM ENGINEERS

In shifting his focus to the viability of an organism's offspring, Tautz implicitly invokes an explanatory and analytic framework that is in one crucial respect directly complementary to that of the genetic paradigm. In effect, through its dependence on mutational analysis, genetics seeks to explain development by asking what causes it to fail or go astray. This approach implicitly assumes that the causes of normal development can be inferred by a kind of logical subtraction—that is, by an enumeration of all those genes that can be identified by their phenotypic failure. By contrast, a focus on developmental stability leads one to ask: What is required to make it work? What is it that endows the developmental process with the reliability required to ensure its survival?

This distinction bears some resemblance to the division between genetics and classical embryology that had prevailed throughout the first half of the twentieth century and that made so many early embryologists wary of the claims of their colleagues in genetics.[24] But it is equally familiar—perhaps even more so—to any engineer attempting to formulate design principles for complex systems predicated on the reliability of performance (for example, the design of airplanes that can be reasonably certain to reach their destination despite the vicissitudes of weather and air traffic they encounter). In short, the aims (or concerns) of engineers led them early on—in fact, long before the advent of information theory—to just that principle of design which has been most inaccessible to the traditional techniques of genetic analysis: if one's aim is to ensure the reliability of a system, the first step is to build in extensive redundancy.[25]

Some of the early embryologists seemed to be familiar with at least one aspect of this general way of thinking. H. Braus, a German embryologist from the turn of the twentieth century, invoked the term "double assurance" in 1906 to characterize the dependence of skeletal structure on the arrangement of muscle tissues.[26] And many years later, Hans Spemann elaborated on this principle in his Silliman Lectures, referring to it as a "synergetical principle of development." Spemann wrote: "The expression 'double assurance' is an engineering term. The cautious engineer makes a construction so strong and durable that it will be able to stand a load which in practice it will never have to bear."[27] But from their war-related efforts of the 1940s, engineers in the middle of the twentieth century learned to look to other ways of ensuring reliability—not only by building in conventional mechanisms of structural redundancy but also by exploiting complex systems of interaction that are, by their very organization, self-stabilizing. These lessons too were soon absorbed into the thinking of at least some embryologists.

C. H. Waddington was one (and almost certainly the best remembered) of the embryologists who so profited. He wrote: "During the recent war, engineers attained some facility in designing machines to carry out tasks which earlier generations would have considered beyond the capacities of anything but an intelligent being . . . The ideas suggested by these self-regulating mechanisms are both very relevant to biology and rather novel."[28] Waddington himself worked in the Operation Research Section of the Royal Air Force Coastal Command, and it was from his own experience and from that of his friends in self-steering gunnery that he learned to draw the analogy which was to become increasingly familiar in the cybernetics of the 1950s and 1960s: "The behaviours of an automatic pilot, of a target-tracking gunsight, or of an embryo, all exhibit the characteristics of an activity guided by a purpose."[29] Indeed, it was at this time, and in this context, that his work on canalization began.[30]

In Waddington's first introduction of the term, he wrote, "The main thesis is that developmental reactions as they occur in organization submitted to natural selection, are, in general, canalized. That is to say, they are adjusted so as to bring about one definite end result regardless of minor variations in conditions during the course of the reaction." In his view, canalization is built into the organism by natural selection as a consequence of its obvious advantages: "It ensures the production of the normal, that is, optimal type in the face of the unavoidable hazards of existence."[31] *Canalization* was a term Waddington had borrowed from his reading of Alfred North Whitehead, and the concept clearly accorded with much of his own prewar thinking about "epigenetic landscapes."[32] But it was only after the war that he began to envision the possibility of a theoretical account of such characteristic features of biological organization. An explanation of "developmental canalization," he wrote, requires supplementing conventional gene theory with an "epigenetic theory"—one in which discrete and separate entities of classical genetics would be displaced by collections of genes which could 'lock in' development through their interactions."[33] In other words, an account of developmental stability needs to be sought in the complex system of reactions that make up the developmental process.

The search for quantitative models displaying such behavior underlay much of Waddington's theoretical efforts well into the 1970s. However, he soon concluded that the particular models developed by Ross Ashby and other cyberneticians on self-organizing systems were not really appropriate to biological development. Instead, he concentrated on feedback models of cross-reacting systems of metabolic reactions. Yet he did not have a great deal of success with these models. Indeed, it is not his theoretical work but the experimental work from his laboratory in the 1940s and '50s that now, with the benefit of hindsight, attracts the most interest.

In a series of studies that were explicitly guided by his focus on developmental stability, Waddington found a way to use the standard techniques of genetics to detect the presence of genetic variation that is phenotypically silent. By selecting for mutants exhibiting greater than usual variability in pattern formation (such as the number of bristles on the thorax of *Drosophila*) and then subjecting these to intense selection pressures, he was able to infer the existence of extensive, though hidden, genetic variability in the wild-type population. Such hidden variability contributes simultaneously to the robustness and the flexibility of development, lending the organism an increased adaptability to unexpected environmental stresses. The fact that this variability is not expressed in wild-type individuals, he argued, shows that the "pattern is in some way stabilized or buffered. The effect of the abnormal gene . . . must have been to produce some destabilization of the pattern, so that the inherent genetic variability could come to expression and be submitted to selection."[34]

Over the last few years, Waddington has begun to enjoy something of a revival among biologists, and it has been suggested that his work on canalization constitutes "a premature discovery."[35] But, as I have already discussed in Chapter 3, Waddington's particular perspective on genetics was far from popular among American geneticists in the 1950s and '60s, and, given its entanglement during that period in debates about the inheritance of acquired characters, the work on canalization could scarcely have avoided being seen as especially problematic.[36] The more significant point for this account, however, is that, rightly or wrongly, Waddington's efforts to formulate an epigenetic theory of development held little interest for most of his colleagues at the time and, despite his many writings on the subject of canalization (and despite the efforts of a number of dedicated students—Brian Goodwin, for example), this work had little impact on the course of research in developmental genetics over the decades that followed.[37]

## New Conjunctions between Engineering Know-How and Biological Wisdom

*Robustness* and *reliability*, along with *complexity*, have become the new buzzwords of the scientific and technical literature of the 1990s, and much of this interest arises out of ever-more-rapid advances in computer technology. In fact, the terms are closely linked, for increases in complexity bring with them increasing opportunities for error and increasing risks of failure. In engineering, therefore, the need for new techniques to manage unreliability and ensure robustness assumes an urgency that grows with each new technological development.

One obvious arena in which such a need arises is in distributed computing systems—best embodied, perhaps, in the Internet. Indeed, as more and more users become interconnected and as our daily lives become more and more dependent on these interconnections, the problem begins to assume mammoth proportions. Leslie B. Lamport defines a distributed computing system as "one in which the failure of a computer you didn't even know existed can render your own computer unusable."[38] Improving the dependabil-

Figure 11: **Marriage of computers and organisms.** *(By Gloria Sharp, for* Los Alamos Science; *reproduced by permission of Los Alamos National Laboratory.)*

ity of individual computers and programs certainly helps, but even if individual computers and programs could be made perfectly dependable, risks of failure would arise from their very interconnectivity. Guaranteeing the robustness of a distributed system therefore requires measures of a different kind. To meet this need, programmers have begun to develop what Birman and van Renesse call "software for reliable networks." These are "programs that allow computer systems to restore normal operation even when problems occur . . . The resulting systems do not need to shut down even if some sites go off-line. Instead they resume service by rapidly reconfiguring to work around crashed servers."[39]

A closely related problem arises from the demand for ever faster and cheaper computational power. Over the last quarter of a century the computational efficiency of silicon-based digital computers has increased by a factor of 10,000, and an increase of another thousand-fold is anticipated by the year 2012. After that, however, current approaches are expected to encounter significant physical limitations, and computer designers are busy looking for alternative design principles for computer architecture. One direction this search is taking is toward the development of quantum computers. Another is to employ logic devices made up of molecules rather than integrated circuits. The latter effort is said to be bottom-up rather than top-down because it relies on the spontaneous synthesis (self-assembly) and chemical interconnectivity (self-ordering) of the constituent units. A chemically assembled computer would be vastly cheaper than present-day computers, but it would also be danger-ously vulnerable to statistical fluctuations in both self-assembly and self-ordering. Thus the question arises: How can one build a reliable computer out of error-prone (and often defective) elements?

James Heath and his colleagues have recently reported some success in building just such a defect-tolerant computer, and on the basis of their success they suggest that "it may be possible to chemically synthesize individual electronic components with less than 100 percent yield, assemble them into systems with appreciable uncertainty in their connectivity, and still create a powerful and reliable data communications network."[40] They define defect tolerance as "the capability of a circuit to operate as desired without physically repairing or removing random mistakes," and they argue that such a capability can be achieved by relegating the work of "repair" to software that incorporates high communication bandwidths, sufficiently dense connectivity, frequent self-testing, and extensive reliance on local rather than global "intelligence."[41] "In a typical microprocessor, a description of what the chip should do is first developed, and then the hardware is constructed on the basis of that logic. The general idea [here] is conceptually the opposite. A generic set of wires, switches, and gates are fabricated in the factory, and then the resources are configured in the field by setting switches linking them together to obtain the desired functionality."[42]

At least some of the lessons learned in the development of defect-tolerant computer architecture echo new principles of design that Rodney Brooks and others have been developing for robust and flexible robotic systems since the mid-1980s. The assumption that had dominated artificial intelligence since the 1950s was that problem-solving ability results from the operation of a core centralized intelligence on a symbolic description (or representation) of the world already inscribed in the system. But after thirty years, the results were noticeably disappointing. Few of the systems designed this way proved able to operate in the real world, and even the most successful examples were too brittle and too inflexible to be of much use. "Artificial Intelligence," wrote Brooks in 1990, "has foundered in a sea of incrementalism."[43]

The alternative he and his colleagues proposed was to design autonomous agents capable of carrying out the tasks they would encounter in their interactions with the world, rather than tasks for which they had been explicitly engineered. How might this be done? The solution that emerged was interactive programming—software designed to sense stimuli from the environment encountered as a result of the actions of the robot and then search for relevant subordinate programs for the processing of the information thus acquired. Brooks called his approach "behavior-based robotics," and he identified the central features of behavior-based robotics as situatedness and embodiment. Rather than dealing with abstract descriptions, robots are "situated in the world," where they deal with "the 'here' and 'now' of the environment that directly influences the behavior of the system." So too, they are embodied, and hence they "experience the world directly—their actions are part of a dynamic with the world, and the actions have immediate feedback on the robots' own sensations."[44]

Brooks and his colleagues implemented these properties in four ways: through the use of (1) parallel circuitry, so that a number of tasks can be executed simultaneously; (2) "behavior-based decomposition," that is, the breaking down of behavior into a number of independently executable sense-act loops; (3) local rules of interaction, whereby the response of a unit depends only on signals from its immediate environment; and (4) principles of robust layered control, in which the various layers can function independently but are so ordered that higher levels can subsume the roles of lower levels when required for the execution of higher order tasks. In a series of stunning successes, they have built robots exhibiting the kind of adaptability and apparent intelligence previously familiar only in the biological world (and perhaps especially characteristic of insect behavior).[45] In these "creatures," problem-solving ability (like functionality in general) requires neither internal representation nor a centralized, preprogrammed capacity for symbol processing. Instead, it appears "as an emergent property of the intensive interaction of the system with its dynamic environment."[46]

The computer science literature on reliability and flexibility abounds with references to biological systems. But it is in my final example that principles of biological organization are most explicitly invoked, and this example comes from an effort under way at MIT that goes by the name *amorphous computing*. Here, however, biology serves simultaneously as inspiration and as domain of application. Gerald Jay Sussman is one of the leaders of this effort, and his starting point is in fact identical to that of Rodney Brooks: "Computer science is in deep trouble. Structured design is a failure. Systems, as currently engineered, are brittle and fragile. They cannot be easily adapted to new situations . . . We need new ideas. We need a new set of engineering principles that can be applied to effectively build flexible, robust, evolvable, and efficient systems."[47] And not surprisingly, it is to biology that Sussman looks for guidance: "From biology we learn that multiple strategies may be implemented in a single organism to achieve a greater collective effectiveness than any single approach. For example, cells maintain multiple metabolic pathways for the synthesis of essential metabolites or for the support of essential processes."[48]

To be sure, engineers have developed powerful strategies for making systems reliable and robust—strategies that rely both on building in redundancy at every level and on programs that compare the output from each pathway or subsystem and, according to some prespecified criteria, select the best of these for the next level of processing. But by drawing inspiration from biology, Sussman asserts, we have learned that it is possible to do even better: "We can provide a mechanism for consistency checking of the intermediate results of the independently designed subsystems, even when no particular value in one subsystem exactly corresponds to a particular value in another subsystem."[49] Such mechanisms lie at the heart of *amorphous computing*, and they depend on descriptions of function and relations between parts that "'let go' of the specific logic of individual processes."

Sussman and his colleagues offer a succinct definition of the challenge of amorphous computing in their recently issued "white paper": "How does one engineer prespecified, coherent behavior from the cooperation of immense numbers of unreliable parts that are interconnected in unknown, irregular, and time-varying ways?"[50] And, following that, they continue with an almost equally succinct description of its research agenda: "In essence, amorphous computing demands new approaches to fault-tolerance. Traditionally, one seeks to obtain correct results despite unreliable parts by introducing redundancy to detect errors and substitute for bad parts. But in the amorphous regime, getting the right answer may be the wrong idea: it seems awkward to describe mechanisms such as embryonic development as producing a 'right' organism by correcting bad parts and broken communications. The real question is how to abstractly structure systems so we get acceptable answers, with high probability, even in the face of unreliability."

As engineers, their first and foremost goal is to put these ideas to practical use, and they envision two ways of doing so: first, by fabricating systems that look to biology "not just as a metaphor, but as the actual implementation technology for a new activity of *cellular computing*," and, second, by developing the control over these processes that will enable them, as engineers, "to make novel organisms with particular desired properties."

Biologists clearly have different aims, but they can nonetheless profit from the insights of computer scientists, and perhaps especially so when the insights of their more technologically oriented colleagues are drawn from their own subject. Indeed, it might be argued that the conceptual traffic between engineering and biological science has never been heavier or more profitable than in the last few years. For example, in their overview of the impact of molecular biology on our current understanding of cellular organization, Hartwell and his colleagues suggest that biologists are now in a position to make good use of—and perhaps even need—assistance from the "synthetic sciences" such as engineering and computer science in their efforts to understand the cellular processes that lead to biological function.[51] The first step is a proper description of these processes. "To describe biological functions," they write, "we need a vocabulary that contains concepts such as amplification, adaptation, robustness, insulation, error correction and coincidence detection."[52]

But more than words, such a description brings with it a certain awareness that is absolutely crucial: namely, the recognition (effectively built into these words) that such properties do not arise from the components of a system (be they individual genes, proteins, or even "modules") but from the interactions among these components.[53] The question, of course, is *how?* And here, too, biologists have found they can profit from the expertise of their colleagues in computer science, making use not only of their vocabulary and general perspective but also of the tools they have helped develop for analyzing such interactions.

Indeed, one of the more conspicuous benefits issuing from the rise of genomics has been the emergence of a new subdiscipline, computational biology, and, in concert, of a new breed of biologists equipped with both biological and computational skills. When the Human Genome Project was first launched in the late 1980s, it was already evident that conventional methods would not suffice for managing the masses of data that a full-scale sequencing effort would yield. Accordingly, a significant portion of the HGP's promotional efforts was directed toward recruiting computer scientists and establishing centers of computational biology (or *bioinformatics*). While initially directed at problems of data management, these new centers provided a home for a wide range of mathematical and computational modeling efforts in biology and fostered the construction of new bridges between experimental and theoretical biology on the one hand and between pure and applied biology on the other. The number of such centers has grown dramatically in the years since, and the collaborations they have spawned are among the major sources of new perspectives in molecular biology, and perhaps especially of the growing recognition of the need to shift to levels of organization higher than that of the gene.

Largely as a consequence of his own studies in metabolic engineering, James Bailey at the Institute of Biotechnology in Zurich takes the "growing indications that in many cases either single genes do not affect phenotype, or that their influence on phenotype does not arise in a simple, obvious fashion" as his starting point and draws what are, to him, the most conspicuous lessons for functional genomics:

These observations and many others support the hypothesis that cells are robust systems that are insensitive to many mutations, particularly those affecting critical "core" activities. The implication of this robustness is failure of many genes, or signals or regulatory interactions, to have any significant effect on phenotype unless a certain set of other genes is simultaneously altered.

Another manifestation of robustness in cellular systems is modulation at several levels to perturbations in gene transcription, evident or strongly suggested in early cloned protein production and metabolic engineering experiments and now reiterated and more globally revealed in genome-wide data sets. These data show that (1) change in the relative level [or concentration] of a particular transcript does not imply a corresponding change in the relative level of the corresponding protein; (2) change in the level of a particular protein does not imply a corresponding change in its in vivo specific activity; and (3) change in the in vitro specific activity of a protein does not imply a corresponding change in the rate of the corresponding reaction or step in the cell. The essential missing concept in assumptions to the contrary is the array of kinetic interactions within the cell's integrated system that determine component, subsystem, and overall organism functional characteristics.[54]

For Bailey, as for an increasing number of others, the moral is clear, and he states it simply: "The current cascade of complete genome sequences, unleashed by microchemical technology and at our fingertips thanks to bioinformatics resources and the Internet, now compels a major shift in bioscience research toward integration and system behavior."[55]

## Where Computers and Organisms Part Ways

As computers and organisms become ever more entangled by the interweaving of ideas, skills, and vocabulary among their home disciplines and, perhaps more bewilderingly, by new modes of material construction and intervention, it becomes difficult at times to know which is serving as a metaphor for the other, or even to distinguish our descriptions of one system from those of the other. Thus, for example, Hartwell and his colleagues review the "design principles of biological systems" that have become familiar to engineers (positive and negative feedback, coincidence detection, amplification, parallel circuitry, fail-safe systems) and then conclude: "Designs such as these are common in biology."[56] At the same time, however, and as the authors clearly recognize, there is one rather conspicuous point at which computers and organisms must definitively part ways, and that, of course, is the route by which the two kinds of systems came by such strikingly similar mechanisms. However much they may have been influenced by biological structures, computers nevertheless are built by human design, while organisms evolve without the benefit of a designer (or so it is generally presumed). The crucial question for biologists is therefore this: By what sort of evolutionary process did such complex self-organized beings come into existence? How can a process dependent solely on the chance appearance of new mutations have given rise to structures whose function is to provide pockets of resistance to the disordering forces of chance—structures designed, that is, to be robust?

But perhaps I am drawing the distinction too sharply. That, in any case, is what Hartwell and his colleagues would seem to imply. They argue that, despite the fact of being man-made, in engineering—just as in biological evolution—"particular solutions in computing, or for any engineered object, are the result of an elaborate historical process of selection by technological, economical and sociological constraints."[57] And one would surely have to agree. Indeed, one might even argue that computers, like organisms, are also selected for their capacities to survive and (in some sense of the term) to reproduce.

Nonetheless, a crucial difference remains. Even if their products are bound by technological constraints and subject to the vicissitudes of social and economic forces, engineers are intelligent designers whose interventions are, by definition, external to their systems. The presumption of most biologists today is that no such comparable agency need be assumed to have been at work over the course of evolutionary history. Yet mechanisms that ensure developmental robustness have nonetheless evolved, and indeed to such a degree of sophistication that even to this day their organizational principles have much to teach engineers. The question we are therefore left with is this: By what kinds of evolutionary processes could such mechanisms have arisen without assistance from human ingenuity?

I am tempted to leave this chapter where it began, with a bow to the wisdom accrued by "a billion years of experimentation," or, as François Jacob might put it, to the generative creativity of eons of *bricolage*—of chance recombinations of existing parts that, by virtue of such recombinations and with the help of ongoing feedback both from their neighbors and from their environment, artlessly acquire new functions.[58] After all, Darwin himself instructed us not to lose sight either of the fundamental historicity of biological function or of the creative potential of historical accumulation, mindless though it may be. As he wrote: "If a man were to make a machine for some special purpose, but were to use old wheels, springs, and pulleys, only slightly altered, the whole machine, with all its parts, might be said to be specially contrived for its present purpose. Thus throughout nature almost every part of each living being has probably served, in a slightly modified condition, for diverse purposes, and has acted in the living machinery of many ancient and distinct forms."[59]

Yet I believe there is more waiting to be said. Darwin taught us the importance of chance in evolution by natural selection, but he also taught us the importance of challenge. In a similar spirit, I suggest that challenge provides a powerful driving force, as well, for the evolution of our understanding of the processes of biological evolution. Already, we can see signs of that evolution in the recent efforts of evolutionary theorists to make sense of the evolved mechanisms for genetic stability, for evolvability, and for developmental robustness that molecular analyses have begun to reveal. Thus I prefer to end with the prediction of a great deal more to come, perhaps even of another Cambrian period, only this time not in the realm of new forms of biological life but in new forms of biological thought.

Conclusion

## WHAT ARE GENES FOR?

Who has not seen *Jurassic Park?* Unless you are one of the very few who has not, you will surely remember that climactic moment when *Tyrannosaurus rex* chases the Jeep in which Ellie and Ian are trying to make their escape. Good theatre for sure, but as Jack Horner has been telling us for years, not very good science. We know from the paleontological evidence that, far from being a menacing predator, *T. rex* was a 12,000-pound animal who could neither run nor grasp, nor, for that matter, see very much in front of his nose. He did, however, have a remarkably well-developed sense of smell, and this is what enabled him to find the rotting carcasses on which he most likely fed. Thus, as Horner and Dobb tell us, if "the Jeep had crashed and Ellie and Ian had died, the tyrannosaur might have sniffed out the site, but only after their bodies had been rotting long enough to broadcast a telltale odor. In any event, the dinosaur would not have pursued the Jeep as it sped away."[1]

Horner is a crusader who has spent much of his life trying to set the record straight on dinosaurs. In fact, Horner's work provided the inspiration for the novel on which *Jurassic Park* was based, and he was also scientific advisor to Steven Spielberg in the making of the film. But after decades of misinformation, setting the record straight is an up-hill struggle. Although Spielberg kept to the paleontological facts as far as his medium and his audience would allow, he knew he must not stray too far from the fanciful image that had become rooted in the popular imagination.

From the early 1900s until 1992, the number-one model from which that image was drawn had dominated the center hall of the American Museum of Natural History in New York City, transfixing the millions of children and adults who, year after year, had come to gaze in awe at the formidable figure of the king of the dinosaurs. There he stood, the "tyrant lizard king," towering above the crowd with his giant tail on the ground, poised to lunge at the next victim on the horizon (see Figure 12).

The truth, however, is that *T. rex* was no lizard but rather closer to an avian. Moreover, as Horner and Dobb explain: "Any specimen display that shows the tail resting on the ground is displaying a broken tail. And to make a dinosaur stand upright while looking straight ahead, the back must be broken as well. So must the neck, in two places. Properly understood, *T. rex*'s skeleton tells us instead that the six-ton animal typically leaned forward, counterbalancing the weight of its immense head and hefty tail over an anatomical pivot—its legs." Finally, "*T. rex*'s arms are so short that they cannot be joined together; they cannot grasp themselves, to say nothing of another animal that happens to be running away."[2]

Because of its many discordances with more recent paleontological data, the AMNH model was finally disman-

Figure 12: The fossil skeleton of *Tyrannosaurus rex*. As originally mounted at the American Museum of Natural History in 1915. *(Slide Number 6218(2); reproduced by permission of the Department of Library Services, American Museum of Natural History.)*

tled in 1992. But even with the new exhibit that has now been constructed (see Figure 13), it's going to take a long time to displace an image that is already so deeply ensconced. And *Jurassic Park* illustrates the difficulty. Many of Spielberg's efforts at portraying a more authentic dinosaur are clearly discernable in the film, yet with that one brief lapse into established stereotype, he tacitly contributes to extending the life of the mythical *T. rex*.

Much the same might be said about popular images of the gene, and even about the many conscientious efforts of biologists to provide us with a more sophisticated understanding of genetic processes. The image of genes as clear and distinct causal agents, constituting the basis of all aspects of organismic life, has become so deeply embedded in both popular and scientific thought that it will take far more than good intentions, diligence, or conceptual critique to dislodge it. So, too, the image of a genetic program—although of more recent vintage—has by now become equally embedded in our ways of thinking, along with its attendant conviction (as Jacob and Monod first put it) that "the genome contains not only a series of blue-prints, but a coordinated program of protein synthesis and the means of controlling its execution."[3] In fact, this very image underlies what, from a scientific perspective, is far and away the most egregious problem of *Jurassic Park*, namely, its utterly fantastic premise that one could clone a dinosaur from its DNA.

The fact of the matter is that talking or writing about genetic processes has become all but impossible without at least sometimes lapsing back into older stereotypes. The very words *gene* and *genetic* are imbued with these earlier meanings—as William Gelbart suggests, genes carry too much "historical baggage." Furthermore, despite the in-

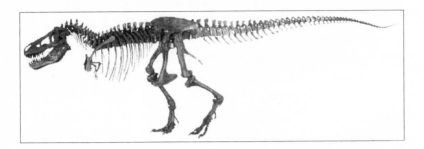

Figure 13: *Tyrannosaurus rex.* As reconstructed in 1992. *(Slide Number K17219; reproduced by permission of the Department of Library Services, American Museum of Natural History.)*

creasing difficulties that biologists face in providing a generally agreed upon and stable definition of a gene, any suggestion that they simply give it up altogether will be seen as both impractical and unrealistic. As I remarked in my Introduction, Johannsen's "little word" has by now become far too entrenched in our vocabulary, and until a new and better vocabulary—and not just a new word—becomes available, biologists will not, indeed they cannot, stop talking about genes.

Furthermore—and here the analogy between genes and dinosaurs breaks down precipitously—the very baggage carried by genes has a richness and breadth of appeal that will be difficult to replicate. But what is probably the most compelling force maintaining talk of genes in biology (at least in the near future) stems from the fact that the word itself is still doing yeoman work, and of a kind that too much has now come to depend upon. I have argued throughout this book that our new understandings of the complexity of developmental dynamics have critically undermined the conceptual adequacy of genes as *causes* of development; further-more, recent developments in molecular biology have given us new appreciation of the magnitude of the gap between genetic information and biological meaning. Thus, from the perspective of the developing organism, the question of what genes are *for* has become increasingly difficult to an-swer. Yet, paradoxically, gene *talk* continues to have obvious and undeniable uses. It might therefore be useful to refor-mulate the question and inquire instead about what gene talk is for.

To shift attention to the question of what gene talk is for requires a substantial enlargement of perspective. We need to take into account not only the context of the bio-logical organism but also the manifestly wider context of biological culture, in both its material and its social as-pects. Before explaining what I mean by this, however, I need to insert a cautionary note, made necessary by widespread misunderstanding about the relation between words and things. Some readers might recall that eminently reassuring slogan from childhood, "Sticks and stones may break my bones, but words will never harm me!" Yet we all know that words *can* harm us, and not simply by hurting our feelings. Just as they can, and indeed do, help us—probably in more ways than we can begin to realize. Words enable us. And while it scarcely needs saying that words are not the same kind of entities as things, that we must not confuse the one with the other, we do quite often need reminding of the equally serious mistake that comes from thinking of things as independent of words. Words have a power to impinge on the world that is unquestionably real. But from what, one might ask, does that power derive? Not from their physi-cality, to be sure; nor from any sort of mystical bond once imagined as tying words to things. Nor, for that matter, does it derive from a direct (or literal) correspondence between a word and a thing. Instead, the power of words derives from a relation to things that is always, and of necessity, mediated by language-speaking actors.

Like the rest of us, scientists are language-speaking actors. The words they use play a crucial (and, more often than not, indispensable) role in motivating them to act, in directing their attention, in framing their questions, and in guiding their experimental efforts. By their words, their very landscapes of possibility are shaped. Thus, to understand how gene talk has affected the course of biological research, we need to examine the particular ways in which terms like *gene, gene action, genetic program* have participated in shaping the biological landscape of the scientists doing that work.

In a sense, this is what I have been doing all along in this book. But I have barely scratched the surface. What is missing—and would be absolutely required for understanding the role of language in biological research—is a far deeper investigation of the material, economic, and social context in which that language functions. Accordingly, I take this Conclusion as an occasion to acknowledge the limitations of my account and also to indicate, if only briefly, the kinds of issues involved in a fuller understanding of the work for which gene talk continues, even now, to be useful.

The first point to be made is obvious to any working biologist (and probably to any working scientist): There may be "no single fact of the matter about what the gene is,"[4] but neither is there necessarily a problem in such a state of affairs. Indeed, the sort of definitional difficulties historians and philosophers worry about rarely if ever impede biologists in their day-to-day usage of the term. But how can this be? Don't scientists require great precision in the language they use? Well, yes and no—that is, in some ways they do, but in other ways, just as in ordinary communication, too much precision would in fact be paralyzing. Where precision is necessary (and absolutely so) is in particular laboratory practices. Moreover, it is from the specificity of the experimental context in which they are invoked that technical terms acquire the precision they need. Terms like *gene* may be subject to a variety of different meanings; but locally, misunderstanding is avoided by the availability of distinct markers directly and unambiguously tied to specific experimental practices. Within that practice, the marker has a clear and unambiguous reference. Change the practice, and different markers will need to be employed. And inevitably, these different markers will pick out somewhat different physical entities.[5] Nevertheless, as long as one stays within the context of a given and clearly understood set of experimental conventions, the term *gene* can still safely serve as an operational shorthand indicating (or pointing to) the marker of immediate experimental significance.

But still, wouldn't it be better—not to mention less prone to confusion down the road—to forsake talk of genes and to use only those words that designate the actual markers? Less ambiguous, perhaps, but definitely not better, and for one simple reason. The meaning of an experimental effect depends on its relation to other effects, and the use of language too closely tied to particular experimental practices would, by its very specificity, render communication across different experimental contexts effectively impossible. Some flexibility in terminology is necessary for the construction of bridges between these different contexts; in turn, such bridges work to guide biologists in their exploration of phenomena that are, by definition, still poorly understood, ill-defined, and open-ended. In other words, the construction of scientific meaning depends on the very possibility of words taking on different meanings in different contexts—that is, it depends on linguistic imprecision. Thus a new set of questions arises: Which differences are to be subsumed under one verbal umbrella? How much imprecision is constructive? When does it become useful to exchange one lexicon, one order of signs, for another? And finally, how is scientific understanding (or meaning) itself transformed by such lexical shifts?[6]

The second point is this: In many cases, experimental markers of the kind discussed above can serve as actual handles—that is, they can be manipulated in such a way as to induce definite and reproducible effects. In the early part of the century, when H. J. Muller imagined the prospect of controlling genetic change in ways that would "place the process of evolution in our hands," his vision was little more than a fantasy. But with the advent of the recombinant DNA revolution, Muller's fantasy has come to look more and more like a realizable prospect. Over the last quarter of the twentieth century, largely as a consequence of that revolution, we have acquired the technical capability to target and alter specific sequences of nucleotides, thereby turning molecular markers that earlier could only be seen into handles for effecting specific kinds of change. Today, with techniques for inducing modifications in the DNA of plants and animals that reliably result in a new or enhanced production of particular proteins, genetic engineering has become a reality. In fact, the efficacy of such interventions is what persuades many molecular biologists of the causal power of genes. As the molecular biologist Robert Weinberg has written, the reason he and his colleagues are convinced that genes are the causal agents of development and that "the invisible agents they study can explain . . . the complexity of life" is that, by manipulating these agents, it is now "possible to change critical elements of the biological blueprint at will."[7]

Is there a problem with Weinberg's inference? Again, yes and no. Certainly, no problem arises if one understands *cause* in the immediate or practical sense of the term as implying nothing more than a way of implementing an effect. But *cause* in such a conspicuously pragmatic sense makes no claim either on generality or on long-term consequences: an inference of the kind Weinberg makes is limited to the specifics of particular experimental interventions and, because of the large number of variables involved, need not even be incompatible with counter-examples. Long ago, the philosopher R. C. Collingwood described such a view of cause as the defining characteristic of what he called the "practical sciences." According to Collingwood, a researcher in the "practical sciences" never tries to give a complete enumeration of causes: "Why should he? If I find that I can get a result by certain means I may be sure that I should not be getting it unless a great many conditions were fulfilled; but so long as I get it I do not mind what these conditions are. If owing to a change in one of them I fail to get it, I still do not want to know what they all are; I only want to know what the one is that has changed."[8]

In just this sense, handles are short-term causes. And for just this reason, demonstrations of their efficacy can offer little assurance to those who continue to worry about the effects of other (perhaps unknown) variables, and about the kinds of unexpected consequences to which such effects can lead over the long-term.[9] In the view of Barbara McClintock, such disregard—such a restrictive sense of causality—is responsible for many of the environmental catastrophes we have encountered, for the ways in which technology, based on the partial analyses of scientists, "is slapping us back in the face very hard": "We were making assumptions we had no right to make. From the point of view of how the whole thing actually worked, we knew how part of it worked . . . We didn't even inquire, didn't even see how the rest was going on. All these other things were happening, and we didn't see it."[10]

Finally, one last function of gene talk must be mentioned, and this concerns its use as a tool for persuasion. Never in the history of the gene has the term had as much force in the popular imagination as in recent years, and, accordingly, never has gene talk had more persuasive—that is, rhetorical—power.[11] The invocation of genes has proven demonstrably effective not only in securing funding and promoting research agendas but also (and perhaps even especially) in marketing the products of a rapidly expanding biotech industry. Indeed, the new partnerships between science and commerce that are daily being forged by the promises of genomics bind genetics to the market with a strength and intimacy that is unprecedented in the annals of basic research in the life sciences. The closer such ties, the greater the research scientist's investment becomes in the rhetorical power of a language that works so well.

Such connections are obvious. But less obvious are the ways in which efficacy as tools of persuasion reinforces the perceived value of gene talk in more immediate experimental contexts, and vice versa. Resonances between and among these different effects make it exceedingly difficult to give up on gene talk, in the laboratory as in the marketplace. One might say that such forms of mutual reinforcement are just what make the terminology of scientific practice self-stabilizing, at least in the short run.

But over the long run, as everybody knows, language does evolve, in science as elsewhere. It always has, and it always will. Indeed, our understanding of the natural world could scarcely progress without such evolution. And while I have just been drawing attention to some of the forces behind the conservation of gene talk, the focus of the main part of the book has been on the shortcomings demanding its transformation. It is only fitting that I close with a reminder of the impetus for change we have seen resulting from developments in research practices—perhaps most dramatically, from the gene's most recent confrontations with advances in molecular genomics. As always, the counter-forces working to destabilize a particular set of terms and concepts emerge out of what might be most simply described as science's ongoing encounter with the real world—from the accumulating inadequacies of an existing lexicon in the face of new experimental findings. And so, I have argued, has been the case with the gene.

Before reaching its limits, the lexicon of the gene had first to be built up, and the history of genetics in the first three quarters of this century offers eloquent testimony to the versatility and strengths of that core concept. Evidence accumulated over the last quarter century, however, provides a different sort of testimony: it shows us that, even in its youth, Johannsen's little word, so innocently conceived in the early days of this century, had had to bear a load that was veritably Herculean. One single entity was taken to be the guarantor of intergenerational stability, the factor responsible for individual traits, and, at the same time, the agent directing the organism's development. Indeed, one might say that no load had seemed too great—as long, that is, as the gene was seen as a quasi-mythical entity. But by the middle part of the century, the gene had come to be recognized as a real physical molecule—in fact, just a bit of DNA—and here, at this point in time, the history of genetics takes its most surprising turn. Both the excitement and the triumph of the new science of molecular biology came from the remarkable evidence it provided suggesting that, incredibly enough, the gene, now understood as a self-replicating molecule of DNA, was a structure equal to its task. Yet, with the maturation of molecular biology, the impracticality (perhaps even impossibility) of that load has become steadily easier to discern.

New kinds of data gathered over the last few decades have dramatically fleshed out our understanding of the parts played by genes in cellular and organismic processes, and in doing so they have made it increasingly apparent how far the weight of such a load exceeds what any one single entity can reasonably be expected to bear, and hence, how appropriate that it be distributed among many different players in the game of life. Indeed, even taking these burdens separately, evolution has apparently seen fit to distribute each of them among a variety of players.

Thus, for example, in Chapter 1, we saw that, by itself, DNA is not capable of guaranteeing its own fidelity from one generation to another—that it needs the help of a complex machinery of editing, proofreading, and repair. Yet more surprisingly, we have seen that such mechanisms not only maintain fidelity but also play an active role in setting the limits of fidelity, by triggering other mechanisms that actively generate genetic variability under conditions of stress. Similarly, in Chapter 2 we looked at a few of the many new phenomena that have vastly complicated not only the early picture of one gene–one trait but also the more recent picture of one gene–one enzyme. We have long known that the rate of protein synthesis requires cellular regulation, but now we have learned that even the question of what kind of proteins are to be synthesized is, in part, answered by the kind and state of the cell in which the DNA finds itself. In higher organisms DNA sequence does not automatically translate into a sequence of amino acids, nor does it, by itself, suffice for telling us just which proteins will be produced in any given cell or at any stage of development. Like the responsibility for maintaining fidelity, this work too is distributed among the many players involved in post-transcriptional regulation. The same can be said regarding the determination of how a protein functions.

Of course, all the protein and RNA molecules participating in such higher-order regulation need themselves to be synthesized and hence must in some sense be "encoded" in the DNA; moreover, awareness of this need is surely what sustains the widespread assumption of a genetic program directing the proceedings. But in Chapter 3, I argued that the assumption of a program inscribed in the DNA also requires rethinking, and I suggested in its place the more dynamic concept of a distributed program in which all the various DNA, RNA, and protein components function alternatively as instructions and as data. Indeed, I argued that the notion of a distributed program accords far better with the picture of cellular regulation and development that has emerged over the last quarter of a century than does the earlier notion of a genetic program.

Finally, in Chapter 4, I explored recent findings of extensive genetic and functional redundancy that fall outside the genetic paradigm and that, in doing so, return us to a problem of central concern to many embryologists in the early part of the century. This is the problem not of genetic stability but of developmental stability—of the conspicuous robustness of developmental processes and their capacity to stay on track despite inevitable environmental, cellular, and even genetic vicissitudes. Can the language of genetics be revised to encompass such effects, or does it need to be supplemented by altogether different concepts and terms? Engineers have developed a conceptual toolkit for the design of systems—like airplanes, for example, or computers—in which reliability is the first and foremost criterion. As such, their approach might be said to be directly complementary to that of geneticists, and I suggest that the latter might profitably borrow some of the concepts and terms developed in the study of dynamic stability to enlarge their own conceptual toolkits.

Where my own sympathies lie should by now be apparent. Genes have had a glorious run in the twentieth century, and they have inspired incomparable and astonishing advances in our understanding of living systems. Indeed, they have carried us to the edge of a new era in biology, one that holds out the promise of even more astonishing advances. But these very advances will necessitate the introduction of other concepts, other terms, and other ways of thinking about biological organization, thereby inevitably loosening the grip that genes have had on the imagination of life scientists these many decades. My hope is that such new concepts and new ways of thinking will soon work to loosen the even more powerful grip that genes have recently come to have on the popular imagination. For if, as Gelbart suggests, the term *gene* may in fact have become a hindrance to the understanding of biologists, it has perhaps become even more of a hindrance to the understanding of lay readers, misleading as often as it informs. As a consequence, it shapes popular hopes and anxieties in ways that are often off target, and in fact counter-productive to effective discussion of public policy even where the issues are real and urgent. At my most optimistic, I even imagine the possibility that new concepts can open innovative ground where scientists and lay persons can think and act together to develop policy that is both politically and scientifically realistic.

# Notes

## Introduction: The Life of a Powerful Word

1 Bateson, W. (1906).

2 Johannsen, W. (1909), p. 124.

3 Johannsen, W. (1911), pp. 132–134.

4 Morgan, T. H. (1933), p. 315.

5 De Vries, H. (1889 [1910]), p. 13.

6 Muller, H. J. (1929).

7 Beadle, G. W., and Tatum, E. L. (1941); Beadle, G. W. (1945); Watson, J. D., and Crick, F. (1953a, 1953b).

8 Watson, J. D. (1992), p. 164.

9 Stephens, C. (1998), p. R47.

10 Gilbert, W. (1992), pp. 84–85.

11 Gelbart, W. (1998), p. 659.

12 See, for example, Lander, E. (1996); Miklos, G. L. G., and Rubin, G. M. (1996); Hieter, P., and Boguski, M. (1997).

13 Hieter, P., and Boguski, M. (1997), p. 601.

14 Ibid.

15 Burtis, K. C., and Hawley, R. S. (1999), p. 125.

## 1. Motors of Stasis and Change: The Regulation of Genetic Stability

1 Darwin, C. (1859), p. 167.

2 Weismann, A. (1885), quoted in Gabriel, M. L., and Fogel, S. (1955), p. 200.

3    De Vries, H. (1889 [1910]), p. 13.

4    Quoted in Portugal, F. H., and Cohen, J. S. (1977), p. 105. For a particularly interesting discussion of Weismann in relation to the increasing distance over the course of the century between genes and the body, see Griesemer, J. R. (forthcoming).

5    Wilson, E. B. (1896), p. 13.

6    Quoted in Portugal, F. H., and Cohen, J. S. (1977), p. 104.

7    Johannsen, W. (1909), p. 124.

8    Wilson, E. B. (1923), p. 280.

9    Quoted in Carlson, E. A. (1971), p. 161. To help frame Muller's thinking, a few historical markers might be helpful: 1896, radioactivity discovered; 1900, rediscovery of Mendelian "factors" (particulate genetic elements); 1901, spontaneous "transmutation of elements" observed by Rutherford and Soddy (so named by Soddy in 1901); 1902, identification of nuclear chromosomes as the site of genetic factors; 1909, coinage of the term *gene*; 1911, Rutherford's discovery of the atomic nucleus; 1919, Rutherford induces first "artificial transmutation."

10   Quoted in Carlson, E. A. (1971), p. 161. Indeed, after Rutherford's success in 1919 in inducing a transmutation of the elements, Muller pursued his own search for a means of inducing mutation with that precedent directly in mind, even entitling his discovery of X-ray induced mutations "Artificial Transmutation of the Gene" (Muller, 1927).

11   Schroedinger, E. (1944), p. 49.

12   Timoféeff-Ressovsky, N. W., Zimmer, K. G., and Delbrück, M. (1935).

13   Schroedinger, E. (1944), p. 61.

14   One of the best accounts (and surely the most extensively documented discussion) of this history is to be found in Olby, R. (1974).

15   Avery, O. T., MacLeod, C. M., and McCarty, M. (1944).

16   Ibid., p. 155.

17   Hershey, A. D., and Chase, M. (1952). It was the use of a Waring

blender to separate the viruses from their host bacteria that lent this experiment its name.

18  Watson, J. D. (1968), p. 197.

19  Stent, G. (1968), p. 390.

20  Ibid., p. 394.

21  See, for example, Radman, M. (1988).

22  Strauss, B. (1995), p. 1511.

23  Hotchkiss, R. (1968), p. 857.

24  Letter to Friedberg, E., February 21, 1995, quoted in Friedberg, E. (1997), p. 17.

25  Frank Stahl to E. F. Keller, September 1, 1997 (e-mail).

26  In response to the question of whether people were surprised, Stahl wrote, "I think (for me) the realization dawned rather slowly, precluding any sense of surprise."

27  Additional mechanisms guaranteeing the integrity of the entire chromosome have also been identified, but these will not be discussed here.

28  To complicate matters even further, evidence is now beginning to emerge suggesting that other enzymes, organized into other repair pathways, may work to monitor and correct errors in transcription, translation, and even in protein structure. Such additional mechanisms would ensure a degree of stability of biological organization going well beyond that of merely genetic stability.

29  Haynes, R. (1988), p. 577.

30  In bacteria, the ratio of harmful to adaptive mutations is estimated at 100,000 to 1.

31  McClintock, B. (1983). Reprinted in Federoff, N., and Botstein, D. (1992).

32  Interview with M. S. Fox by Leslie Barber, November 19, 1998. Although the initial experiment was never published, an improved version, performed in Radman's lab, was reported some years later (Defais, M., Caillet-Fauquet, P., Fox, M. S., and Radman, M., 1976).

33 See Witkin, E. (1989), p. 32.

34 Quoted in Friedberg, E. (1997), p. 277.

35 Radman, M. (1974), p. 134. Today, however, Radman interprets this early proposal more radically—that is, as predicting the existence of certain enzymes that would specifically enhance diversity and adaptability in the endangered population by allowing individual cells to mutate when their survival is under threat; see Radman, M. (1999), p. 866.

36 Indeed, bacterial mutants that lack key components of the SOS system fail to show the usual increase in mutagenesis expected under exposure to ultraviolet light.

37 Taddei, F., Vulić, M., Radman, M., and Matić, I. (1997).

38 Taddei, F., Radman, M., Maynard Smith, J., Toupance, B., Gouyon, P. H., and Godell, B. (1997).

39 Sniegowski, P. D., et al. (1997).

40 Radman, M., Matić, I., and Taddei, I. (1999).

41 The distinction between organism and population is blurred, however, by the fact that the effect of such inducible (or transient) mutagenesis is to alter the organism. Thus here too the evolution of such mechanisms requires the evocation of some kind of group selection, operating either on the level of the population of genes in a given genome or on the level of a population of organisms.

42 Radman, M. (1999), p. 866.

43 Ibid.

44 Caporale, L. H. (1999).

45 Shapiro, J. A. (1999), p. 32.

46 Dawkins, R. (1988).

47 Ibid., p. 201.

48 Ibid., p. 218.

49 Alberch, P. (1991).

50 Gerhart, J., and Kirschner, M. (1997), p. 613.

51 Kirschner, M., and Gerhart, J. (1998), p. 8427.

52 Delbrück, M. (1949).

53   For further remarks on the stability of cellular inheritance, see Chapter 3, n. 32.

54   Nanney, D. L. (1957), p. 134.

## 2. The Meaning of Gene Function: What Does a Gene Do?

1   My narration of this history is manifestly a gloss, and historians of genetics will be particularly sensitive to its neglect of the extensive debates among classical geneticists about the nature, structure, and function of the gene that not only complicate the story given here but would be absolutely essential to a full historical account. The aim of this sketch is different: it is to identify the landmarks of classical genetics that are of most immediate relevance (and hence that provide the backdrop) to the initial triumphs and subsequent development of molecular genetics.

2   Morgan, T. H. (1928), p. 27.

3   See Keller, E. F. (1995), chap. 1.

4   Schroedinger (1944), p. 23.

5   See, for example, Baltimore, D. (1984), p. 150.

6   De Vries, H. (1889 [1910]), p. 4.

7   1911, quoted in Olby, R. (1976), p. 145.

8   Muller, H. J. (1951), p. 95.

9   Muller, H. J. (1936), p. 214.

10   De Vries, H. (1910), p. 194.

11   One of the best such accounts is to be found in Olby, R. (1976).

12   Troland, L. T. (1917), p. 328.

13   Haldane, J. S. (1931), p. 147. J. S. Haldane is now best known as the father of the geneticist J. B. S. Haldane, but in his own time (1860–1936) he was widely regarded as Britain's leading physiologist. For further discussion, see Keller, E. F. (2000c).

14   Watson, J. D., and Crick, F. (1953b), p. 96. See Chapter 1 for a discussion of the importance of this argument for considerations about genetic stability.

15    Crick, F. (1957), p. 152.

16    Reverse translation was ruled out in Crick's formulation of the "Central Dogma" (1957).

17    Rheinberger, H. J. (1996), p. 10.

18    Jacob, F., and Monod, J. (1959).

19    Jacob, F., and Monod, J. (1961a), pp. 197–198.

20    Monod, J., and Jacob, F. (1961), p. 394.

21    Morgan, T. H. (1934), p. 9.

22    For an account of the historical context of Monod and Jacob's work (as well as for a somewhat different reading of its significance), see, for example, Gaudillière, J. P. (1988, 1993); Griesemer, J. R. (2000).

23    Jacob, F., and Monod, J. (1959), p. 1282.

24    Part of the reason the answer to this question is not obvious is that the terminology varies so much in its usage. For example, the response to a query I recently sent to the yeast genome web site reads, "The 6200 estimate includes all ORFs (structural + regulatory genes)." An "ORF" (or "open reading frame") is usually defined as "a DNA sequence that is uninterrupted by a stop codon and encodes part or all of a protein." When I asked for clarification, I was directed to another web site giving more detail (*http://www.mips.biochem.mpg.de/proj/yeast/tables/inventy.html*). But this site too leaves the question unanswered, for, while counting by ORF's would normally not include sequences used only as templates for RNA molecules, some RNA molecules (for example, transfer RNA and snRNA—short RNA transcripts that associate with proteins to form small nuclear ribonucleoprotein particles participating in RNA processing) are included in the data base.

25    Although credit for the original discovery of split genes in 1977 belongs primarily to Richard Roberts and Phillip Sharp, it was Walter Gilbert (1978) who coined the terms *intron* and *exon*. As Gilbert put it, "The gene is a mosaic: expressed sequences held in a matrix of silent DNA, an intronic matrix."

26 The amount of intronic DNA varies greatly, but in some eukaryotic genes it is as high as 95 percent.

27 See, for example, the discussion of alternative splicing in Alberts, B., et al. (1994).

28 The first indication that "junk" DNA may serve functional roles came as early as 1982, but recognition of its importance in normal gene function has grown rapidly with the rise of genomics.

29 For example, a recent review article reports the identification of 576 possible splicing variants of a gene (*cSlo*) active in the hair cells of the inner ear of the chick. The gene *cSlo* encodes a protein that plays a crucial role in determining the resonant frequency of the cell, and variations in the sequence of that protein alter its responses to different sound frequencies—in effect, "tuning" the chick's ear to incoming sounds. Many (although not yet all) of these variants have been observed, and it is thought that homologous genes in humans and mice might encode an even greater number of possible splice variants. But, asks Douglas Black, "What are the instruments?" "Who plays the melody?" and, finally, "Who is the conductor?" See Black, D. L. (1998).

30 For example, *ras* and GAP are key proteins in a signal transduction pathway regulating mammalian growth. But, asks Alan Hall, "Who's Controlling Whom?" "Does *ras* control GAP, or does GAP control *ras*? The answer," he concludes, "seems to be: yes." See Hall, A. (1990), p. 923.

31 In their discussion of RNA editing, G. M. Malacinski and D. Freifelder (1998), p. 329, write: "It remains to be determined if this co- or posttranscriptional modification takes place in the nucleus or the cytoplasm."

32 Monod, J., Changeaux, J. P., and Jacob, F. (1963). For an excellent historical analysis of this work, see Creager, A. N. H., and Gaudillière, J. P. (1996).

33 Jeffery, C. J. (1999), p. 453.

34 Burian, R. M. (1985), p. 37.

35 Portin, P. (1993), p. 208.

36 Gelbart, W. (1998), p. 660.

37 In fact, the very category of "genetic disease" has expanded so greatly as to now include conditions (such as cancer) that are only very rarely genetic in the usual sense of the term, that is, in the sense of being passed on through the germ line.

38 Weatherall, D. J. (1997), p. 4.

39 Rheinberger, H. J. (1995).

40 The distinction I am making here between the "structural gene" and the "functional gene" roughly corresponds to the distinction between the "evolutionary gene" and the "molecular gene" advocated by Paul Griffiths and Eva Neumann-Held (1999).

41 A useful introduction to these issues from a philosophical perspective can be found in Sterelny, K., and Griffiths, P. E. (1999).

42 See, for example, Strohmann, R. (1997); Duboule, D. (1997).

43 Pattee's original question was, "How does a molecule become a message?" See Pattee, H. (1969).

44 Johannsen, W. (1911), p. 132.

### 3. THE CONCEPT OF A GENETIC PROGRAM: HOW TO MAKE AN ORGANISM

1 Sturtevant, A. H. (1932), p. 304.

2 Ibid., p. 307.

3 First introduced in Mayr, E. (1959).

4 Beermann, W. (1956), p. 222.

5 Waddington, C. H. (1954), pp. 114–115.

6 See the discussion of Delbrück's analysis of multiple steady states in Chapter 1. Waddington, C. H. (1954), pp. 115–116.

7 See Chapter 4, n. 36.

8 Waddington, C. H. (1948).

9 Although Waddington had been an active experimentalist throughout the 1930s, '40s, and even into the '50s, by the end of that decade his attention had largely turned to the revival of a "theoretical biology." In the late 1960s he organized a series of

"Symposia on Theoretical Biology" under the sponsorship of the International Union of Biological Sciences (Waddington, C. H., 1968–1972).

10 Jacob, F., and Monod, J. (1961b), p. 354.

11 Simultaneously, and almost surely independently, Ernst Mayr introduced the notion of "program" in his 1961 article on "Cause and Effect in Biology" (adapted from a lecture given at MIT on February 1, 1961). There he wrote, "The complete individualistic and yet also species-specific DNA code of every zygote (fertilized egg cell), which controls the development of the central and peripheral nervous system . . . is the *program* for the behavior computer of this individual" (Mayr, E., 1961, p. 1504).

12 Jacob, F. ([1970] 1976), p. 2.

13 Ibid., p. 4.

14 Ibid., pp. 8–9.

15 Ibid., p. 9.

16 See Atlan, H., and Koppel, M. (1990), for a discussion of the distinction between "program" and "data."

17 Apter, M. J., and Wolpert, L. (1965), p. 257. See also Apter, M. J. (1966), and, for a more extended discussion, Keller, E. F. (1995), chap. 3, and Keller, E. F. (2000b).

18 See, for example, Wolpert, L., and Lewis, J. H. (1975).

19 Bonner, J. (1965), p. 6.

20 Ibid., p. v.

21 Ibid., p. 6.

22 Ibid. (In this quote, I have silently corrected "property" to "properly.")

23 Ibid., p. 6.

24 Ibid., p. 134.

25 Ibid., p. 135.

26 Ibid.

27 Another way of saying this is to describe the end-point of development as an adult organism capable of reproducing (or participating in the reproduction of) its own starting point.

28 Goldfarb, D. S. (1997).

29  Bonner himself writes of the roles played by chromosomal proteins (histones), hormones, and RNA molecules; today, the list has expanded considerably to include enzymatic networks, metabolic networks, transcription complexes, signal transduction pathways, and so on, with many of these additional factors embodying their own "switches."

30  The terminology employed in this research is somewhat slippery, and I put the words *cloning* and *nuclear transfer* in quotes to alert the reader of this fact. As defined in *The Chambers Dictionary* (1993, Chambers Harrap), a "clone" is "a group of two or more individuals with identical genetic makeup derived, by asexual reproduction, from a single common parent or ancestor." But as it is used in recent literature on cloning, "cloning" has come to refer to any animals produced by nuclear transfer, whether from a single parent or not, and whether or not fully identical in genetic makeup to the parental donor of the nucleus (differences may, for example, inhere in the different mitochondrial genes contributed by the two parents). For this reason, Keith Campbell has suggested that "The resultant animals may therefore be more aptly described as 'genomic copies'" (Campbell, K., 1999, p. 245). Similarly, "nuclear transfer" does not always imply the removal of the nucleus from one cell and its transfer into another, enucleated, cell: in fact, the cloning of Dolly was achieved not by nuclear transfer in the strict sense of the term but by the fusion of a complete adult cell with an oöplast (see Keller, E. F., and Ahouse, J. C., 1997).

31  Wilmut, I., et al. (1997). Readers will of course recall that the birth of Dolly brought more than acclaim; it also aroused intense fears in readers everywhere as a portent of the possibilities and the dangers of cloning in humans. The best discussion of these issues I know of is to be found in Lewontin, R. C. (1997).

32  Delbrück's steady state model (discussed in Chapter 1) is an example of just such a mechanism (Delbrück, M., 1949). In the years since, other such mechanisms have been proposed as well. But until quite recently, it was generally supposed that such

epigenetic mechanisms of inheritance could not pass through the germ line. Today, however, a number of examples of epigenetic modifications persisting through the germ line have also been reported, and the number of such examples is growing rapidly; see, for example, Cubas, P., Vincent C., and Coen, E. (1999); Morgan, H. D., Sutherland, H. G. E., Martin, D. I. K., and Whitelaw, E. (1999). For a review of the burgeoning literature on epigenetic inheritance up to 1995, see Jablonka, E., and Lamb, E. (1995).

33  Briggs, R., and King, T. J. (1952).

34  For a review of this work, see Gurdon, J., et al. (1979).

35  Wilmut, I., et al. (1997).

36  Stewart, C. (1997), p. 769.

37  The term *chromatin* was introduced by the German cytologist Walther Flemming in 1880 to denote the nuclear material that showed up under staining during the period prior to cell division when the chromosomes organize and contract, and for a long time, the chromatin was widely taken to be equivalent to the genetic material. Indeed, as late as 1968, the term is defined in a standard glossary as "that part of the nuclear material that makes up the genetic material and contains the genetic information of the cell" (Rieger, R., Michaelis, A., and Green, M. M., 1968, p. 63). This history has in fact contributed greatly to the ambiguity of the term *genetic program,* for it encourages a certain degree of slippage between "genetic" and "chromosomal" and hence, blurs the distinction I am making between genetic and developmental programs (see Keller, E. F., 2000a). Because the protein components of chromatin structure play so critical a role in the regulation of gene expression, and because they are not themselves genetic, I count them as part of the developmental program. In other words, the distinction I am drawing between genetic and developmental programs cuts quite differently from older debates over the relative importance of nuclear and cytoplasmic factors and is notably more in line with current distinctions between genetic and epigenetic.

38  Campbell, K. (1999), p. 250.

39  Ibid., p. 250.

40  Brenner, S., et al. (1990), p. 485. For further discussion of the evolution of Brenner's views of genetic programs, see de Chadarevian, S. (1998).

41  Wade, N. (1998), p. 1.

42  Ibid.

43  Garcia-Bellido, A. (1998), pp. 112–113.

44  Davidson, E., et al. (1998), p. 1896.

45  Ibid., p. 1902.

46  See my discussion in Chapter 2.

47  Halder, G., Callaerts, P., and Gehring, W. (1995). Calling this gene *eyeless* now seems decidedly inappropriate, but, like many other genes, it was first identified by the failure to develop a fully formed eye in its mutant form.

48  Gehring, W. (1998), p. 204. Gehring's estimate of 2,500 genes required to make a *Drosophila* eye is impressive, for that number may correspond to as much as 30 percent of the total number of genes in *Drosophila* (currently estimated at somewhere between 8,000 and 17,000). Direct evidence for the importance of background conditions (genetic and otherwise) comes from the fact that expression of the homologous gene does not lead to the formation of an eye in all tissues.

49  Darwin, C. (1859), p. 186.

50  Coen, E. (1999), pp. 87–88.

51  Ibid., p. 1.

52  Supplementing Lenny Moss's observation that a genetic program is "an object nowhere to be found" (Moss, L., 1992, p. 335), I would argue that the "program" for gene expression (that is, the developmental program) is everywhere to be found.

53  Holland, P. W. H. (1999).

54  Leland Hartwell and his colleagues summarize the history of our understanding of biological function as follows: "Much of twentieth-century biology has been an attempt to reduce biological phenomena to the behaviour of molecules. This approach is

particularly clear in genetics, which began as an investigation into the inheritance of variation . . . From these studies, geneticists inferred the existence of genes and many of their properties, such as their linear arrangement along the length of a chromosome. Further analysis led to the principles that each gene controls the synthesis of one protein, that DNA contains genetic information, and that the genetic code links the sequence of DNA to the structure of proteins. Despite the enormous success of this approach, a discrete biological function can only rarely be attributed to an individual molecule . . . In contrast, most biological functions arise from interactions among many components" (Hartwell, L. H., et al., 1999, p. C47).

## 4. LIMITS OF GENETIC ANALYSIS: WHAT KEEPS DEVELOPMENT ON TRACK?

1   Delbrück, M. (1949).

2   Gould, S. J. (1989).

3   To the extent that organisms participate in the construction of their own niches, they can also be said to exert a stabilizing control over their most immediately relevant environment. My discussion of developmental stability in this chapter could therefore be usefully augmented by a discussion of the evolution of mechanisms for ensuring niche stability. For this, however, I refer the reader to Avital, A., and Jablonka, E. (2000).

4   A similar view can be found in Maeshiro, T., and Kimura, M. (1999), where the authors write: "The assumption that the robustness and changeability are prerequisites for the survival and evolution of organisms is applicable to all aspects of evolution . . . The requirements for robustness and changeability are perhaps the single most universal aspect underlying the evolution of life."

5   See, for example, Gerhart, J., and Kirschner, M. (1997).

6   By Treviranus and Oken in Germany, and by Lamarck in France. In point of fact, however, the same term had already been in-

voked two years earlier in England, in a marginal note in a medical treatise by Burdach (see Schiller, J., 1978, p. 1).

7    Kant, I. (1993), 66, p. 558.

8    Ibid.

9    Ibid., 65, p. 557 (italics in original).

10   Ibid.

11   "Il y a comme un dessin préétabli de chaque être et de chaque organe, en sorte que si, considéré isolément, chaque phénomène de l'économie est tributaire des forces générales de la nature, pris dans ses rapports avec les autres, il révèle un lien special, il semble dirigé par quelque guide invisible dans la route qu'il suit et amené dans la place qu'il occupe" (Bernard, 1878, p. 51, quoted in Jacob, F., 1976).

12   Ibid., p. 4.

13   Weaver, W. (1949), p. 540.

14   See, for example, Andrew Hodges's biography of Alan Turing (1983).

15   See especially von Neumann's 1949 lecture, printed in von Neumann, J. (1966). For further discussion of postwar preoccupations with "self-organization," see Keller, E. F. (2000b).

16   Jacob, F. (1976), p. 9.

17   See Morange, M. (1998), chap. 7.

18   · Angier, N. (1993).

19   The observation of high degrees of genetic polymorphism lent strong support to Motoo Kimura's "neutral theory of evolution" and is now thought by some to be linked to the more recent observations from "knockout" experiments.

20   Brenner, S., et al., (1990).

21   Tautz, D. (1992), p. 263.

22   Thomas, J. H. (1993), p. 395.

23   Tautz, D. (1992), p. 264.

24   See, for example, Keller, E. F. (1995), chap. 1.

25   The importance of an engineering approach to biology has in fact been forcefully argued by the philosopher of biology William Wimsatt for many years (see, for example, Wimsatt, W.,

1981; 2000). Inspired by von Neumann's article on building reliable organisms from unreliable components (1956), Wimsatt has urged biologists and philosophers of biology to think of robustness as an evolutionary design principle.

26 Braus, H. (1906).

27 Spemann, H. (1938), pp. 92–93.

28 Waddington, C. H. (1957), p. 141.

29 Waddington, C. H. (1971), p. 20.

30 Edward Yoxen, however, has failed to find any evidence of a specific influence of Waddington's wartime work on the formulation of his thinking about canalization (see Yoxen, E. J., 1986).

31 Waddington, C. H. (1942).

32 See, for example, Gilbert, S. F. (1991), p. 199.

33 Waddington, C. H. (1948).

34 Waddington, C. H. (1962), p. 226.

35 Wilkins, A. (1997), p. 257.

36 It should be noted that Waddington's influence was considerably greater among British geneticists than among American geneticists, and at least part of the reason for this is surely political. Although it is not the purpose of this book to examine the social and political context of genetics research in the twentieth century, no discussion of the reception of Waddington's work on canalization and "genetic assimilation" can omit mention of its inevitable association with the arguments of Lysenko, and of the particularly costly effects of such an association in the Cold War era (see, for example, Gilbert, S. F., 1991, p. 205).

37 Brian Goodwin may be the most visible of Waddington's students, but despite his persistent arguments that an understanding of morphogenesis requires a more global conceptual framework than that of genetics (see, Goodwin, B., 1985), and, more specifically, that morphogenesis is an intrinsically robust process (Goodwin, B., et al., 1993), his influence has been largely confined to the conspicuously marginal world of "theoretical biology."

38 Quoted in Birman, K. P., and van Renesse, R. (1996), p. 48.

39    Ibid., p. 50. According to Birman and van Renesse, the principal techniques employed in the development of fault-tolerant software are threefold: "active replication" (in which a system's software makes redundant copies of vital programs or servers as they are used); "load-sharing" (parceling out data among servers); and "modularity" (in which different modules can be fit together in various combinations to support specific needs).

40    Heath, J. R., Kuekes, P. J., Snider, G. S., and Williams, R. S. (1998).

41    The analogy invoked by Heath and his colleagues to illustrate the last point is the difference between American and Japanese postal systems: "If residences are laid out in a Cartesian coordinate system, then it does not take much complexity in the mail-delivery system to find an address. In Japan, however, there are no regular street addresses. Nevertheless, the knowledge of many local postmen is sufficient to deliver a letter" (ibid., p. 1720).

42    Ibid., p. 1717.

43    Brooks, R. (1990), p. 3.

44    Brooks, R. (1991), p. 1227.

45    Indeed, Brooks has given the title "Cambrian Intelligence" to his recently published collection of papers on the early history of the new AI (1999). A particularly clear account of "interactive programming" can be found in Lynn Stein's forthcoming book on this subject.

46    Maes, P. (1991), p. 1.

47    Sussman, G. J. (1999).

48    Ibid.

49    Ibid.

50    Abelson, H., et al. (1999).

51    Hartwell, L. H., et al. (1999), p. C47.

52    Ibid.

53    As Hartwell et al. (1999) write, "Cell biology is in transition from a science that was preoccupied with assigning functions to individual proteins or genes, to one that is now trying to cope with

the complex sets of molecules that interact to form functional modules" (p. C53).

54  Bailey, J. E. (1999), p. 616.

55  Ibid., p. 617.

56  Hartwell, L. H., et al. (1999), p. C51.

57  Ibid., p. C52.

58  Jacob, F. (1982).

59  Darwin, C. (1862), pp. 283–284.

### Conclusion: What Are Genes For?

1  Horner, J. R., and Dobb, E. (1997), p. 7.

2  Ibid., p. 5.

3  Jacob, F., and Monod, J. (1961b), p. 354.

4  Burian, R. M. (1985).

5  For example, in one context, the term may refer to only to regions of DNA characterized as an "ORF of length x," while in another, it might include noncoding regions of DNA that are used as templates for RNA molecules. In yet a third context, it might refer to the mature (postsplicing) RNA molecule used in the actual translation process.

6  These might be said to be T. S. Kuhn's questions, for they lie at the heart of his notion of a "paradigm shift." But it was especially the last question that preoccupied T. S. Kuhn over the last decades of his life. A partial sketch of his deliberations on this matter can be found in his Afterword to *World Changes* (Horwich, P., ed., 1994); a far more fully developed account is currently in preparation for posthumous publication.

7  Weinberg, R. A. (1985), p. 48.

8  Collingwood, R. C. (1940), p. 303.

9  Precisely such differences in perspective underlie much of the current controversy over genetically modified foods. Concerns about the safety of gene therapy, only underscored by the recent reports of unforeseen casualties, have a similar basis.

10  Barbara McClintock, quoted in Keller, E. F. (1983), pp. 205–206.

11  Here, a second misunderstanding (and closely related to that discussed above concerning the relation between words and things) also needs addressing. Traditionally, words make up the subject of the humanities, where things define the subjects of the natural sciences. More specifically, the analysis of how language works is the subject of *rhetoric*. But to many scientists, the very word has an aura of disreputability. Often tacitly (if not openly) coupled with the modifier "just," *rhetoric* is widely associated with deceit, and hence is seen as fundamentally antithetical to science. Such a view, however, bespeaks a serious amnesia about the complex and multipurpose ways in which scientific language not only does function, but also, and inescapably, has perforce to function in the real world of human actors and human interests.

# References

Abelson, H., Allen, D., Coore, D., Hanson, C. P., Homsy, G., Knight, T. F., Jr., Nagpal, R., Rauch, E., Sussman, G. J., and Weiss, R. 1999. Amorphous computing. White Paper, MIT. http://www.swiss.ai.mit.edu/projects/amorphous/workshop-sept-99/hal-tk-gjs.pdf.

Alberch, P. 1991. From genes to phenotype: dynamical systems and evolvability. *Genetica* 84:5–11.

Alberts, B., Bray, D., Lewis, J., Raff, M., Roberts, K., and Watson, J. D. 1994. *Molecular Biology of the Cell*. New York: Garland.

Angier, N. 1993. When a vital gene is missing, understudies fill in. *New York Times*, Sept. 7, 1993, p. C3.

Apter, M. J. 1966. *Cybernetics and Development*. Oxford: Pergamon.

Apter, M. J., and Wolpert, L. 1965. Cybernetics and development. *J. Theor. Biol.* 8:244–257.

Atlan, H., and Koppel, M. 1990. The cellular computer DNA: program or data. *Bulletin of Math. Biol.* 52(3):335–348.

Avery, O. T., MacLeod, C. M., and McCarty, M. 1944. Studies on the chemical transformation of pneumococcal types. *J. Exp. Med.* 79:137–158.

Avital E., and Jablonka E. 2000. *Animal Traditions: Behavioural Inheritance in Evolution*. Cambridge: Cambridge University Press.

Bailey, J. E. 1999. Lessons from metabolic engineering for functional genomics and drug discovery. *Nature Biotech.* 17:616–618.

Baltimore, D. 1984. The brain of a cell. *Science* 84:150.

Bateson, W. 1906. The progress of genetic research. In *Third Conference on Hybridization and Plant Breeding*, pp. 90–97. London.

Beadle, G. W. 1945. The genetic control of biochemical reactions. *Harvey Lectures* 40:179–194.

Beadle, G. W., and Tatum, E. L. 1941. Genetic control of biochemical reactions in Neurospora. *Proc. Natl. Acad. Sci.* 21:499–506.

Beermann, W. 1956. Nuclear differentiation and functional morphology of chromosomes. *Cold Spring Harbor Symp. Quant. Biol.* 21:217–232.

Birman, K. P., and van Renesse, R. 1996. Software for reliable networks. *Scientific American*, May, pp. 48–51.

Black, D. L. 1998. Splicing in the inner ear: a familiar tune, but what are the instruments? *Neuron* 20:165–168.

Boguski, M. S. 1999. Biosequence exegesis. *Science* 286:453–455.

Bonner, J. 1965. *The Molecular Biology of Development.* Oxford: Oxford University Press.

Braus, H. 1906. Ist die Bildung des Skeletts von den Muskelanlagen abhängig? *Morph. Jahrb.* 35:38–119.

Brenner, S., Dove, W., Herskowitz, I., and Thomas, R. 1990. Genes and development: molecular and logical themes. *Genetics* 126:479–486.

Briggs, R., and King, T. J. 1952. Transplantation of living nuclei from blastula cells into enucleated frogs' eggs. *Proc. Natl. Acad. Sci.* 38:455–463.

Brooks, R. A. 1990. Elephants don't play chess. *Robotics and Autonomous Systems* 6:3–13.

——1991. New approaches to robotics. *Science* 253:1227–1232.

———1999. *Cambrian Intelligence*. Cambridge: MIT Press.

Burian, R. M. 1985. On conceptual change in biology: the case of the gene. In *Evolution at a Crossroads: The New Biology and the New Philosophy of Science*, ed. D. J. Depew and B. H. Weber, pp. 21–42. Cambridge: MIT Press.

Burtis, K. C., and Hawley, R. S. 1999. The millennium flies in. *Nature* 401:125–126.

Campbell, K. 1999. Nuclear transfer in farm animal species. *Sem. Cell & Dev. Biology* 10:245–253.

Canguilhem, G. 1994. *A Vital Rationalist*. Cambridge: MIT Press.

Caporale, L. H. 1999. Chance favors the prepared genome. In *Molecular Strategies in Biological Evolution*, ed. L. H. Caporale. New York: New York Academy of Sciences.

Carlson, E. A. 1971. An unacknowledged founding of molecular biology: H. J. Muller's contributions to gene theory. *J. Hist. Biology* 4:149–170.

Coen, E. 1999. *The Art of Genes: How Organisms Make Themselves*. Oxford: Oxford University Press.

Collingwood, R. G. 1940. *An Essay on Metaphysics*. Oxford: Clarendon Press.

Creager, A. N. H., and Gaudillière, J. P. 1996. Meanings in search of experiments and vice-versa: the invention of allosteric regulation in Paris and Berkeley, 1959-1968. *Historical Studies in the Physical and Biological Sciences* 27:1–89.

Crick, F. 1957. On protein synthesis. *Symp. Soc. Exp. Biol.* 12:138–163.

Cubas, P., V. C., and Coen, E. 1999. An epigenetic mutation responsible for natural variation in floral symmetry. *Nature* 410:157–161.

Darwin, C. 1859. *On the Origin of Species*. Fac. ed. Cambridge: Harvard University Press, 1964.

———1862. *On the Various Contrivances by Which British and Foreign Orchids*

*Are Fertilised by Insects, and on the Good Effects of Intercrossing.* London: John Murray.

Dawkins, R. 1988. The evolution of evolvability. In *Artificial Life,* ed. C. G. Langton, pp. 201–220. New York: Addison Wesley.

De Chadarevian, S. 1998. Of worms and programmes: "Caenorhabditis elegans" and the study of development. *Studies in Hist. and Phil. of Biol. and Biomed. Sci.* 29:81–105.

De Vries, H. 1889 [1910]. *Intracellular Pangenesis.* Chicago: Open Court.

Defais, M., Caillet-Fauquet, P., Fox, M. S., and Radman, M. 1976. Induction kinetics of mutagenic DNA repair activity in *E. coli* following ultraviolet irradiation. *Molec. Gen. Genet.* 148:125–130.

Delbrück, M. 1948. A physicist looks at biology. In *Phage and the Origins of Molecular Biology,* ed. J. Cairns et al., pp. 9–22. Cold Spring Harbor: Cold Spring Harbor Laboratory Press.

———1949. Discussion. In *Unités Biologiques Douées de Continuité Génétique.* Lyon: Editions du CNRS, 33. (Trans. by Delbrück, ms. sent to Leo Szilard, Feb. 1, 1960.)

Duboule, D. 1997. The evolution of genomics. *Science* 279:555.

Dyson, F. 1985. *Origins of Life.* New York: Cambridge University Press.

Friedberg, E. 1997. *Correcting the Blueprint of Life: An Historical Account of DNA Repair Mechanisms.* Cold Spring Harbor: Cold Spring Harbor Laboratory Press.

Gabriel, M. L., and Fogel, S., eds. 1955. *Great Experiments in Biology.* New York: Prentice-Hall.

Garcia-Bellido, A. 1998. Discussion of S. Brenner, "Biological Computation." In *The Limits of Reductionism,* ed. G. R. Bock and J. A. Goode, pp. 106–116. Novartis Symposium. Chichester: John Wiley.

Gaudillière, J. P. 1988. Un code moléculaire pour la différenciation cellulaire: la controverse sur les transferts d'ARN informationnel (1955-1973) et les étapes de diffusion du paradigme de la biologie moléculaire. *Fundamenta Scientiae* 9:429-467.

——1993. Molecular biology in the French tradition? Redefining local traditions and disciplinary patterns. *J. Hist. Biol.* 26:473-498.

Gehring, W. 1998. *Master Control Genes in Development and Evolution.* New Haven: Yale University Press.

Gelbart, W. 1998. Data bases in genomic research. *Science* 282:660.

Gerhart, J., and Kirschner, M. 1997. *Cells, Embryos, and Evolution.* Oxford: Blackwell Scientific.

Gilbert, S. F. 1991. Induction and the origins of developmental genetics. In *A Conceptual History of Modern Embryology,* ed. S. F. Gilbert, pp. 181-206. New York: Plenum.

Gilbert, W. 1978. Why genes in pieces? *Nature* 271:501.

——1992. Vision of the grail. In *The Code of Codes,* ed. D. J. Kevles and L. Hood, pp. 83-97.

Goldfarb, D. S. 1997. Whose finger is on the switch? *Science* 276:1814-1817.

Goodwin, B. 1985. What are the causes of morphogenesis? *BioEssays* 3(1):32-36.

Goodwin, B., Kauffman, S., and Murray, J. D. 1993. Is morphogenesis an intrinsically robust process? *J. Theor. Biol.* 163:135-144.

Gould, S. J. 1989. *Wonderful Life: The Burgess Shale and the Nature of History.* New York: Norton.

Griesemer, J. R. 2000. Reproduction and the reduction of genetics. In *The Concept of the Gene in Development and Evolution: Historical and Epistemological Perspectives,* ed. P. Beurton, R. Falk, and H-J. Rheinberger. Studies in Philosophy and Biology. Cambridge: Cambridge University Press.

Griesemer, J. R. Forthcoming. The informational gene and the substantial body: on the generalization of evolutionary theory by abstraction. In *Varieties of Idealization,* ed. N. Cartwright and M. Jones. Poznan Studies. Amsterdam: Rodopi.

Griffiths, P. E., and Neumann-Held, E. M. 1999. The many faces of the gene. *Bioscience* 49(8):656–662.

Gurdon, J., Laskey, R. A., De Robertis, E. M., and Partington, G. A. 1979. Reprogramming of transplanted nuclei in *amphibia.* In *International Review of Cytology Supplement 9 Nuclear Transplantation,* ed. J. F. Danielli and M. A. DiBernardino, pp. 161–178. New York: Academic Press.

Hagedoorn, Arend. 1911. Autocatalytic substances: the determinants for the inheritable characters. *Vorträge und Aufsätze über Entwicklungsmechanik der Organismen* (Leipzig), Hft. 12: 1–35.

Haldane, J. S. 1931. *The Philosophical Basis of Biology.* Garden City: Doubleday, Doran.

Halder, G., Callaerts, P., and Gehring, W. 1995. Induction of ectopic eyes by targeted expression of the *eyeless* gene in *Drosophila. Science* 267:1788–1792.

Hall, Alan. 1990. Ras and GAP: who's controlling whom? *Cell* 61:921–923.

Hartwell, L. H., Hopfield, J. J., Leibler, S., and Murray, A. W. 1999. From molecular biology to modular cell biology. *Nature Suppl.* 402:C47.

Haynes, R. 1988. Biological context of DNA repair. In *Mechanisms and Consequences of DNA Damage Processing,* pp. 577–584. UCLA symposium held at Taos, N.M., January 24–30. New York: Alan R. Liss.

Heath, J. R., Kuekes, P. J., Snider, G. S., and Williams, R. S. 1998. A defect-tolerant computer architecture: opportunities for nanotechnology. *Science* 280:1716–1721.

Hershey, A. D., and Chase, M. 1952. Independent functions of viral proteins and nucleic acid in growth of bacteriophage. *J. Gen. Physiology* 36:39–56.

Hieter, P., and Boguski, M. 1997. Functional genomics: it's all how you read it. *Science* 278:601–602.

Hodges, Alan. 1983. *Alan Turing: The Enigma of Intelligence.* London: Burnett Books.

Holland, P. W. H. 1999. The future of evolutionary developmental biology. *Nature Suppl.* 402:C41.

Horner, J. R., and Dobb, E. 1997. *Dinosaur Lives.* New York: Harcourt Brace.

Horwich, P., ed. 1994. *World Changes: Thomas Kuhn and the Nature of Science.* Cambridge: MIT Press.

Hotchkiss, R. 1968. Metabolism and growth of gene substance. *Cold Spring Harbor Symp. Quant. Biol.* 33:857–870.

Jablonka, E., and Lamb, M. 1995. *Epigenetic Inheritance and Evolution.* New York: Oxford University Press.

Jacob, F. 1976 [1970]. *The Logic of Life.* New York: Pantheon.

——1982. *The Possible and the Actual.* New York: Pantheon.

Jacob, François, and Jacques Monod. 1959. Gènes de structure et gènes de régulation dans la biosynthése des protéins. *C. R. Acad. Sci. Paris* 349:1282–1284.

——1961a. On the regulation of gene activity. *Cold Spring Harbor Symp. Quant. Biol.* 26:193–211.

——1961b. Genetic regulatory mechanisms in the synthesis of proteins. *J. Molec. Biol.* 3:318–356.

Jeffery, C. J. 1999. Moonlighting proteins [talking point]. *Trends in Biochem. Sci.* 24(1):8–11.

Johannsen, W. 1909. *Elemente der Exakten Erblichkeitslehre.* Jena: Gustav Fischer.

——1911. The genotype conception of heredity. *Am. Nat.* 45:129–159.

Kant, I. 1790. *Critique of Judgement.* Rpt. *Great Books* 39:461–475, trans. J. C. Meredith. Chicago: Encyl. Brittanica, 1993.

Keller, E. F. 1983. *A Feeling for the Organism.* New York: Freeman.

——1995. *Refiguring Life: Metaphors of Twentieth-Century Biology.* New York: Columbia University Press.

——2000a. Decoding the genetic program. In *The Concept of the Gene in Development and Evolution: Historical and Epistemological Perspectives,* ed. P. Beurton, R. Falk, and H-J. Rheinberger. Studies in Philosophy and Biology. Cambridge: Cambridge University Press.

——2000b. Marrying the pre-modern to the post-Modern: computers and organisms after WWII. In *Growing Explanations,* ed. N. Wise.

——2000c. Is there an organism in this text? In *Controlling Our Destinies: Historical, Philosophical, Ethical, and Theological Perspectives on the Human Genome Project,* ed. P. R. Sloan. Notre Dame: University of Notre Dame Press, pp. 273–290.

Keller, E. F., and Ahouse, J. C. 1997. Writing and reading about "Dolly." *Bioessays* 19(8):741–742.

Kevles, D. J., and Hood, L., eds. 1992. *The Code of Codes.* Cambridge: Harvard University Press.

Kirschner, M., and Gerhart, J. 1998. Evolvability. *Proc. Natl. Acad. Sci.* 95:8420–8427.

Lander, E. 1996. The new genomics: global views of biology. *Science* 274:536–539.

Lewontin, R. C. 1997. The confusion over cloning. *New York Review of Books,* Oct. 23.

Maes, P., ed. 1991. *Designing Autonomous Agents.* Cambridge: MIT Press.

Maeshiro, T., and Kimura, M. 1998. The role of robustness and changeability on the origin and evolution of genetic codes. *Proc. Natl. Acad. Sci.* 95:5088–5093.

Malacinski, G. M., and D. Freifelde. 1998. *Essentials of Molecular Biology* (3rd ed.). Boston: Jones and Bartlett.

Mann, T. 1924 [1952]. *The Magic Mountain.* New York: Knopf.

Mayr, E. 1959. Where are we? *Cold Spring Harbor Symp. Quant. Biol.* 24:1–14.

——1961. Cause and effect in biology. *Science* 134:1501–1506.

Miklos, G. L. G., and Rubin, G. M. 1996. The role of the genome project in determining gene function: insights from model organisms. *Cell* 86:521–529.

Monod, J., Changeux, J. P., and Jacob, F. 1963. Allosteric proteins and cellular control systems. *J. Mol. Biol.* 6:306–329.

Monod, J., and Jacob, F. 1961. General conclusions: teleonomic mechanisms in cellular metabolism, growth, and differentiation. *Cold Spring Harbor Symp. Quant. Biol.* 26:389–401.

Morange, M. 1998. *La Part des Gènes.* Paris: Editions Odile Jacob.

Morgan, H. D., Sutherland H. G. E., Martin, D. I. K., and Whitelaw, E. 1999. Epigenetic inheritance at the agouti locus in the mouse. *Nature Genetics* 23(3):314–318.

Morgan, T. H. 1928. *The Theory of the Gene.* New Haven: Yale University Press.

——1933. The relation of genetics to physiology and medicine. In *Nobel Lectures . . . Physiology and Medicine, 1922–1941,* pp. 313–328. Amsterdam, 1963.

——1934. *Embryology and Genetics.* New York: Columbia University Press.

Moss, Lenny. 1992. A kernel of truth? On the reality of the genetic program. *Phil. Sci. Assn.* 1:335–348.

Muller, H. J. 1927. Artificial transmutation of the gene. *Science* 66:84–87.

——1929. The gene as the basis of life. First presented before the International Congress of Plant Sciences, Section of Genetics, Symposium on "The Gene," Ithaca, N. Y., August 19, 1926; published in *Proceedings of the International Congress of Plant Science* 1:897–921.

———1936. Physics in the attack on the fundamental problems of genetics. *Scientific Monthly* 44:210–214.

———1951. The development of the gene theory. In *Genetics in the Twentieth Century*, ed. L. C. Dunn. New York: Macmillan.

Nanney, D. L. 1957. The role of the cytoplasm in heredity. In *The Chemical Basis of Heredity*, ed. W. D. McElroy and B. Glass, pp. 134–163. Baltimore: Johns Hopkins Press.

Olby, R. 1974. *The Path to the Double Helix*. Seattle: University of Washington Press.

Pattee, H. 1969. How does a molecule become a message? *Dev. Biol. Suppl.* 3:1–16.

Portin, P. 1993. The concept of the gene: short history and present status. *Quart. Rev. Biol.* 68:173–223.

Portugal, F. H., and Cohen, J. S. 1977. *A Century of DNA*. Cambridge: MIT Press.

Radman, M. 1973. Phenomenology of an inducible mutagenic DNA repair pathway in *Escherichia coli:* SOS repair hypothesis. In *Molecular and Environmental Aspects of Mutagenesis,* ed. L. Prakash et al., pp. 128–142. Springfield: Charles C. Thomas.

———1988. The high fidelity of DNA duplication. *Scientific American,* August, pp. 40–46.

———1999. Mutation: enzymes of evolutionary change. *Nature* 401:866–869.

Radman, M., Matic, I., and Taddei, F. 1999. The evolution of evolvability. In *Molecular Strategies in Biological Evolution,* ed. L. H. Caparole. *Annals of N. Y. Acad. of Science* 870:146–155.

Rheinberger, H-J. 1995. Genes: a disunified view from the perspective of molecular biology. *Gene Concepts and Evolution*, MPIWG preprint 18.

———1996. Gene concepts: fragments from the perspective of molecular biology. *Gene Concepts in Development and Evolution*, MPIWG preprint 123.

Rieger, R., Michaelis, A., and Green, M. M. 1968. *A Glossary of Genetics and Cytogenetics*. New York: Springer-Verlag.

Schiller, J. 1978. *La notion d'organisation dans l'histoire de la biologie*. Paris: Maloine.

Schroedinger, E. 1944. *What Is Life?* Cambridge: Cambridge University Press.

Shapiro, J. A. 1999. Genome system architecture and natural genetic engineering. In *Molecular Strategies in Biological Evolution*, ed. L. H. Caparole. *Annals of the N.Y, Acad. Sci.* 870:23-25.

Sniegowski, P. D., Gerrish, P. J., and Lenski, R. E. 1997. Evolution of high mutation rates in experimental populations of *E. coli*. *Nature* 387:703-705.

Spemann, H. 1938. *Embryonic Development and Induction*. New Haven: Yale University Press.

Stein, L. A. Forthcoming. *Interactive Programming*. New York: Morgan Kaufmann Publishers.

Stephens, C. 1998. Bacterial sporulation: a question of commitment? *Curr. Biol.* (B44) 8(2):R45-48.

Sterelny, K., and Griffiths, P. E. 1999. *Sex and Death: An Introduction to the Philosophy of Biology*. Chicago: University of Chicago Press.

Stewart, C. 1997. Nuclear transplantation: an udder way of making lambs. *Nature* 385:769.

Strauss, B. 1995. Molecular pathologies. *Science* 270:1511-1513.

Strohmann, R. C. 1997. The coming Kuhnian revolution in biology. *Nature Biotechnology* 15:194-200.

Sturtevant, A. H. 1932. The use of mosaics in the study of the developmental effects of genes. *Proceedings of the Sixth Int. Cong. of Genetics*, p. 304.

Sussman, G. J. 1999. Robust design through diversity. http://www.swiss.ai.mit.edu/projects/amorphous/workshop-sept-99/robust-diversity.pdf.

Taddei, F., Radman, M., Maynard Smith, J., Toupance, B., Gouyon, P. H., and Godell, B. 1997. Role of mutator alleles in adaptive evolution. *Nature* 387:700–702.

Taddei, F., Vuliæ, M., Radman, M., and Matiæ, I. 1997. Genetic variability and adaptation to stress. In *Environmental Stress, Adaptation and Evolution*, ed. R. Bijlsma and V. Loescheke. Basel: Birkhäuser.

Tautz, D. 1992. Redundancies, development and the flow of information. *BioEssays* 14(4):263–266.

Thomas, J. H. 1993. Thinking about genetic redundancy. *Trends in Genetics* 9(11):395–399.

Timoféeff-Ressovsky, N. W., Zimmer, K. G., and Delbrück, M. 1935. Über die Natur der Benmutation und der Genstruktur. *Nachr. Ges. Wiss. Göttingen, math-phys. Kl.* Fachgr. G(1):189–245.

Troland, L. T. 1917. Biological enigmas and the theory of enzyme action. *Am. Nat.* 51:321–350.

von Neumann, J. 1956. Probabalistic logics and the synthesis of reliable organisms from unreliable components. In *Automata Studies*, ed. C. E. Shannon and J. McCarthy. Princeton: Princeton University Press.

———1966. *Theory of Self-Reproducing Automata*, ed. A. W. Burks. Urbana: University of Illinois Press.

Waddington, C. H. 1942. Canalization of development and the inheritance of acquired characters. *Nature* 150:563–565.

——1948. The genetic control of development. *Symp. Soc. Exp. Biol.* 2:145-154, New York: Academic Press.

——1954. The cell physiology of early development. In *Recent Developments in Cell Physiology*, ed. J. A. Kitching. London: Butterworth.

——1957. *Strategy of the Genes*, London: Allen & Unwin.

——1962. *New Patterns in Genetics and Development.* New York: Columbia University Press.

——1968-1972. *Towards a Theoretical Biology : An IUBS Symposium*, pp. 1-4. Edinburgh: Edinburgh University Press.

——1971. *Biology, Purpose and Ethics.* Worcester, MA: Clark University Press.

Wade, N. 1998. Animal's Genetic Program Decoded, in a Science First. *The New York Times*, p. 1.

Watson, J. D. 1968. *The Double Helix.* New York: Atheneum.

——1992. A personal view of the project. In *The Code of Codes*, ed. D. J. Kevles, and L. Hood, pp. 164-173.

Watson, J. D., and Crick, F. 1953a. A structure for deoxyribose nucleic acid. *Nature* 171:737-738.

——1953b. Genetical Implications of the Structure of Deoxyribonucleic Acid. Nature 171:964-967.

Weatherall, D. J. 1998. How much has genetics helped? *Times Literary Supplement*, January 30, pp. 4-5.

Weaver, W. 1949. Problems of organized complexity. *American Scientist* 36:143-156.

Weinberg, R. A. 1985. The molecules of life. *Scientific American* 253(4):48-57.

Weismann, A. 1885 [1889]. *Continuity of the Germ Plasm.* In *Essays upon heredity and kindred biological problems*, ed. E. Poulton et al. Oxford: Clarendon Press.

Wilkins, A. 1997. Canalization: a molecular genetic perspective. *BioEssays* 19(3):257–262.

Wilmut, I., Schnieke, A. E., McWhir, J., Kind, A. J., and Campbell, K. H. S. 1997. Viable offspring derived from fetal and adult mammalian cells. *Nature* 385:810–813.

Wilson, E. B. 1896. *The Cell in Heredity and Development.* New York: Macmillan.

——1923. The physical basis of life. *Science* 42(1471):277–286.

Wimsatt, W. C. 1981. Robustness, reliability and overdetermination. In *Scientific Inquiry and the Social Sciences,* ed. M. Brewer and B. Collins, pp. 124–163. San Francisco: Jossey-Bass.

——2000. *Re-Engineering Philosophy for Limited Beings: Piecewise Approximations to Reality.* Cambridge: Harvard University Press, forthcoming.

Witkin, E. 1989. Ultraviolet mutagenesis and the SOS response in Escherischia coli: a personal perspective. *Environmental and Molecular Mutagenesis* 14(S16):30–34.

Wolpert, L., and Lewis, J. H. 1975. Towards a theory of development. *Federation Proceedings* 34(1):14–20.

Yoxen, E. 1986. Form and strategy in biology: reflections on the career of C. H. Waddington. In *A History of Embryology: The Eighth Symposium of the British Society for Developmental Biology,* ed. T. J. Horder, J. A. Witkowski, and C. C. Wylie. Cambridge: Cambridge University Press.

Yuh, C-H., Bolouri, H., and Davidson, E. H. 1998. Genomic cis-regulatory logic: experimental and computational analysis of a sea urchin gene. *Science* 279:1896–1902.

## Acknowledgments

I want to thank the many friends and colleagues who have helped in the writing of this book with their critical readings of drafts of chapters, with their willingness to respond to my numerous technical questions, and above all with their general support and encouragement. The accessibility, wisdom, and expertise of my brother, Maurice Fox, were invaluable. I also benefited from every one of the suggestions and criticisms offered by Jeremy Ahouse, Sunny Auyang, Anya Goodman, Jim Griesemer, Eva Jablonka, Jehane Kuhn, and Lynn Stein. But especially I want to thank Loup Verlet, whose support and encouragement never flagged and whose generosity as a reader seems to know no limit. Finally, I want to thank Michael Fisher for his enthusiasm, my illustrator, Nick Thorkelson, for his patience and good humor, and my editor, Susan Wallace Boehmer, for her unfailing good sense.

# Index